The Physics of Atmospheres, Third Edition

In the third edition of *The Physics of Atmospheres*, John Houghton has revised his acclaimed textbook to bring it completely up-to-date. The first two editions received excellent reviews, and the new edition will again prove to be invaluable as a text for students of atmosphere science.

The Physics of Atmospheres provides a comprehensive and concise description of the physical processes governing the structure and the circulation of the atmosphere. Simple physical models are constructed by applying the principles of classical thermodynamics, radiative transfer and fluid mechanics, together with analytical and numerical techniques. These models are applied to real planetary atmospheres. In this new edition, chapters have been introduced on topics of strong contemporary interest such as chaos and atmospheric predictability and climate and climate change. The chapters on global observation (especially through remote sensing) and numerical modelling have also been substantially extended, and all parts of the book have been updated.

Like its predecessors, this new edition will be an essential textbook for advanced undergraduate and graduate courses in meteorology, atmospheric physics, remote sensing, climate science, environmental science, and planetary science. Researchers and professionals in atmospheric physics and meteorology will also find it a state-of-the-art review of their subjects.

Sir John Houghton CBE FRS is Co-chairman of the Scientific Assessment Working Group of the Intergovernmental Panel on Climate Change. He was Professor of Atmospheric Physics at the University of Oxford from 1976–83, Chief Executive of the Meteorological Office from 1983 to his retirement in 1991, and Chairman of the Royal Commission on Environmental Pollution from 1992–98. He has Honorary Doctorates from the Universities of Wales, East Anglia, Leeds, Birmingham, Stirling, Reading, and Heriot-Watt. He has received gold medals from the Royal Meteorological Society and the Royal Astronomical Society and the prestigious International Meteorological Organization Prize. He is an Honorary Member of the Royal Meteorological Society and the American Meteorological Society. He has authored a number of books including *Global Warming: The Complete Briefing* (2nd edn, Cambridge University Press, 1997), which was runner-up for the Sir Peter Kent Conservation Book Prize.

To Janet and Peter

The works of the Lord are great, sought out of all them
that have pleasure therein.
Psalm 111, v. 2

The Physics

of Atmospheres

THIRD EDITION

JOHN HOUGHTON

*Co-chairman of the Scientific Assessment Working Group of
the Intergovernmental Panel on Climate Change*

PUBLISHED BY THE PRESS SYNDICATE OF THE UNIVERSITY OF CAMBRIDGE
The Pitt Building, Trumpington Street, Cambridge, United Kingdom

CAMBRIDGE UNIVERSITY PRESS
The Edinburgh Building, Cambridge CB2 2RU, UK
40 West 20th Street, New York, NY 10011-4211, USA
477 Williamstown Road, Port Melbourne, VIC 3207, Australia
Ruiz de Alarcón 13, 28014 Madrid, Spain
Dock House, The Waterfront, Cape Town 8001, South Africa

http://www.cambridge.org

First published 2002

Printed in the United Kingdom at the University Press, Cambridge

Typeface Times Roman 11.25/14 pt.   *System* QuarkXPress [BTS]

A catalog record for this book is available from the British Library.

Library of Congress Cataloging in Publication Data
Houghton, John Theodore.
    Physics of atmospheres / John Houghton. – 3rd ed.
        p.   cm.
    Includes bibliographical references and index.
    ISBN 0-521-80456-6 – ISBN 0-521-01122-1 (pb.)
        1. Atmospheric physics.   2. Dynamic meteorology.   I. Title.
    QC880 .H68   2001
    551.5′01′53–dc21
                                                                2001025803

ISBN   0 521 80456 6   hardback
ISBN   0 521 01122 1   paperback

Contents

Preface to the first edition

During the last ten years or so, three important factors have combined to bring about a large increase of interest in atmospheric science. First, man's increasing industrial activity has brought into much sharper focus the problems of pollution and the possibility of artificial modification of the environment either inadvertently or in a controlled way. Secondly, concern about world food resources in the face of rapidly increasing population has made us much more aware of the critical effect of fluctuations in climate, particularly in parts of the developing world. Thirdly, the advent of the artificial satellite and developments in electronic computers have made available much more powerful tools for atmospheric research and have made it possible to study the atmosphere as a whole.

At the present time, study of the global atmosphere is being concentrated by a large international programme – the Global Atmospheric Research Programme – which is being organized jointly by atmospheric scientists (through the International Council of Scientific Unions) and the national weather services (through the World Meteorological Organization). The aims of this programme are to attack some of the basic problems regarding our understanding of the behaviour and circulation of the whole atmosphere (hence the term global) and of the mechanisms of climatic change.

It is in the context of this global programme that this book has been written. Its aim is to introduce physics students at both undergraduate and graduate levels to the physical processes which govern the structure and circulation of a planetary atmosphere. In writing the book I have concentrated on the basic physics and wherever possible have constructed simple physical models by applying to the atmosphere the principles of classical thermodynamics, radiation transfer and fluid mechanics.

The book is meant to be a working text and for that reason includes a large number of problems. Some of the problems extend considerably the basic material in the chapters; other problems are included to illustrate what is already in the text.

I have attempted to introduce all the main areas of study involved in the physics of the neutral atmosphere (the ionized atmosphere and the magnetosphere have been deliberately omitted) while at the same time keeping the book within reasonably small compass. This has necessitated a rigorous selection of material. I hope, however, I have achieved a balanced text from which the reader will be able to acquire a perspective of the atmosphere as a whole and a feeling for where the subject is going. A substantial bibliography is included to help the student to pursue further reading. Also included are a number of appendices summarizing information concerning the atmosphere which research workers will find useful for reference.

Because atmospheric science cuts across a number of traditional disciplines in physics and chemistry, symbols, nomenclature and units can be even more of a problem than usual. On the question of symbols I have tended to use those familiar in the various subjects rather than use unfamiliar ones in the interests of uniformity. I have, in general, employed SI units, with one notable exception. For pressure I have kept to the use of millibars, which are universally used in operational meteorology, rather than change to Pascals.

I am indebted to many colleagues and students from whom I have received help in writing the book. Some of the ideas and material for the dynamical chapters arose from an admirable series of lectures on Atmospheric Dynamics given in Oxford by Dr R. S. Harwood, who also made valuable comments and criticisms regarding the text of those chapters. Other criticisms and suggestions on various parts of the text were made by Dr C. D. Walshaw, Dr G. D. Peskett, Professor R. Hide, Professor P. A. Sheppard and by graduate students of Oxford University who have used the text and done the problems. Dr C. D. Rodgers, Mrs M. Corney and Mr R. J. Wells assisted in the collection of some of the material for the appendices, Mr M. J. Wale has assisted with providing the answers to problems and hints to their solutions. Mr D. Wrigley has read through the proofs and provided the index. I also wish particularly to thank Miss C. M. Wagstaff who typed the manuscript, Miss B. Linder who drew most of the diagrams, and the staff of Cambridge University Press for their courtesy and assistance during the preparation of the book.

March 1976   J. T. Houghton

Preface to the second edition

The opportunity of a new edition has been taken to make significant additions and alterations to some of the chapters. In particular, chapter 4 on radiative transfer has been restructured to give a more logical development and chapters 10, 11, 12 and 13 on the general circulation, numerical modelling, observations and climate have been significantly extended to take account of modern developments and increasing interest in these topics.

The inclusion of a large number of problems as a didactic aid has been much appreciated. In this new edition nearly thirty new problems have been added, many of them containing a substantial amount of exposition in order to illustrate principles or to introduce new ideas.

In order to come into line with recommended practice, the unit of pressure throughout this new edition is the Pascal.

I am indebted to many colleagues in the Meteorological Office and elsewhere who have suggested improvements for the new edition and assisted in its preparation. I acknowledge particular assistance from Dr D. G. Andrews, Dr K. A. Browning, Dr M. Cullen, Dr R. Hide and Dr F. W. Taylor who have provided suggestions for (and criticism of) the text or the illustrative material and from Mrs Brenda Bell who typed much of the additional text.

January 1986

J. T. Houghton

Preface to the third edition

Since the second edition was published in 1986, very large scientific and political interest has developed worldwide in the possibility of anthropogenic climate change on a global scale due in particular to the increasing emissions of 'greenhouse gases', for instance carbon dioxide. Much new research has been undertaken on establishing the likely future magnitude and pattern of such climate change and on the degree of predictability that might be achieved in such projections. In this third edition I have therefore added two substantially new chapters – on predictability and on climate change – in which I have addressed some of the basic physics involved in answering these questions.

Because of the developments in observing systems and in numerical modelling, I have also substantially rewritten the chapters (11 and 12) on these topics. Minor revisions and additions have also been made to the earlier chapters. For instance, the theory of the 'Chapman' layer has been added to chapter 5 and cloud radiation feedback (one of the most important climate feedbacks) has been addressed briefly in chapter 6.

I am indebted to many colleagues who have assisted with the preparation of this new edition. Paul Morris updated the material in appendices 9 and 10 and John Barnett the material for fig. 5.1 and appendix 5. Material and diagrams on numerical modelling and climate change were provided by Alan Dickinson, David Carson, Andrew Lorenc, Frank Saunders and Paul van der Linden. Material for the observations chapter was provided by Malcolm Kitchen, David Pick and Bob Shearman. I am also indebted to David Andrews, Michael Cullen, Raymond Hide, Brian Hoskins, Paul Mason, John Mitchell, Tim Palmer, Keith Shine, Fred Taylor and Lance Thomas, who have provided suggestions for the text or the illustrations or provided comments on the draft text. Finally, it is a pleasure to acknowledge the courtesy and assistance of Matt Lloyd and the staff of Cambridge University Press in the preparation of the book.

December 2001   John Houghton

Acknowledgements

Permission has been received to use published diagrams as a basis for figures in the text from the following: Professors J. Charney, H. U. Dütsch, R. Hide, A. S. Monin and E. Palmén; Drs M. Berenger, K. A. Browning, N. Gerbier, R. Hanel, L. Hembree, P. R. Julian, R. A. McClatchey, S. Manabe, P. J. Mason, L. T. Matveev, C. W. Newton, C. Ridley, D. Rodgers, B. Schlachman, J. E. A. Selby, W. L. Smith, F. W. Taylor, L. W. Uccellini, D. Vanous, W. M. Washington and R. T. Wetherald; Academic Press Inc.; Akademie-Verlag, Berlin; Air Force Geophysics Laboratory; American Meteorological Society; D. Reidel Publishing Co.; Eumetsat; Intergovernmental Panel on Climate Change; Israel Programme for Scientific Translations; McGraw-Hill Book Co.; Meteorological Office, National Aeronautics and Space Administration, U.S.A.; National Oceanographic and Atmospheric Administration, U.S.A.; Optical Society of America; Royal Meteorological Society; Taylor & Francis Ltd.; University of Dundee Electronics Laboratory; and the World Meteorological Organization.

1

Some basic ideas

1.1   Planetary atmospheres

The atmosphere of a planet is the gaseous envelope surrounding it. Large differences exist between the atmospheres of different planets both in chemical composition and physical structure as will be seen from the information about the atmospheres of our nearest planetary neighbours contained in table 1.1.

For a complete understanding of an atmosphere we need to know about its evolution and the processes which have determined its mass and its composition. We also need to know about its physical structure and the distribution of density, composition and motion within the atmosphere.

In the case of the earth's atmosphere very detailed information is required in order to predict its state for periods of days or weeks ahead. Further, because of our concern about climatic change, it is necessary to understand the factors which determine the average state of the atmosphere over periods of years and centuries.

In this book discussion will be confined to a description of the physical processes involved in atmospheric study, with application mostly to the earth's atmosphere. Using basic physical principles in thermodynamics, radiation transfer and fluid dynamics, we shall first develop simple models which may be used to give some understanding of the main features of atmospheric structure and behaviour. In later chapters, methods of observing the atmosphere will be described and an indication given of the major problems which are being tackled in atmospheric science at the present time.

1.2   Equilibrium temperatures

A crude estimate of the effective temperature T_e of a planet's surface may be made by equating the solar radiation it absorbs to the infrared radiation it emits as follows:

Table 1.1

	Mean surface temperature (K)	Surface pressure (atm)[a]	Accn due to gravity (m s^{-2})	Main constituents	
Venus	750	90	8.84	CO_2 96% N_2 4%	Deep clouds complete cover
Earth	280	1	9.81	N_2 78%[b] O_2 21%	~50% cover H_2O clouds
Mars	218	0.006	3.76	CO_2 95% N_2 3%	Some very thin H_2O clouds
Jupiter	134[c]	2[c]	26	H_2, He	NH_3 clouds

[a] 1 atm is mean pressure at earth's surface = 1.013×10^5 Pa(N m^{-2}) = 101.3 kPa = 1013 millibars (mb).
[b] See appendix 3 for detailed composition of earth's atmosphere.
[c] At cloud top.

Table 1.2

	R	A	T_e(K)	T_m(K)	M_r	Period of rotation (days)
Venus	0.72	0.77	227	230	44	243
Earth	1.00	0.30	256	250	28.8	1.00
Mars	1.52	0.15	216	220	44	1.03
Jupiter	5.20	0.58	98	130	2	0.41

$$4\pi a^2 \sigma T_e^4 = \pi a^2 (1 - A) F / R^2 \tag{1.1}$$

where σ is the Stefan–Boltzmann constant, a is the radius of the planet at distance R (astronomical units) from the sun, F (=1370 W m^{-2}) is the solar flux on a surface near the earth (i.e. when $R = 1$) normal to the solar beam and A is the *albedo*, i.e. the ratio of reflected to incident solar energy for the whole planet. The right hand side of (1.1) is the absorbed solar radiation and the left hand side the radiation emitted by the planet assuming it behaves as a black body at temperature T_e. Table 1.2 lists values of the various parameters in (1.1) for some of the planets and compares approximate measured temperatures T_m with those calculated from (1.1). The agreement is good except in the case of Jupiter, for which planet absorption of solar radiation accounts for only about half the energy input required to maintain its observed temperature; the other half must, therefore, be internally generated (cf. problem 1.2).

Fig. 1.1. Illustrating the greenhouse effect for a planetary atmosphere that is more transparent to solar radiation than to infrared radiation. Infrared radiation emitted by the planetary surface is absorbed by greenhouse gases (such as water vapour, carbon dioxide, methane) and by clouds. The glass in a greenhouse possesses similar optical properties in that it is transparent to solar radiation and opaque to infrared. Although known as the 'greenhouse effect' it might be noted that in practice it only provides a minor contribution to the warmth of a greenhouse that is largely due to the suppression of air motions within it.

solar radiation

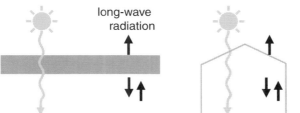

The effective temperature for Venus listed in table 1.2 is very different from the temperature at the planet's surface listed in table 1.1. The very dense atmosphere and complete cloud cover of Venus is substantially opaque to radiation of all wavelengths shorter than about 1 mm. These clouds act as a radiation blanket preventing almost all radiation emitted from the lower atmosphere from escaping, while allowing a small amount of solar radiation through (see §1.5). This process whereby a high surface temperature can be maintained is known as the *greenhouse effect* (fig. 1.1). To a more limited extent than on Venus it is also effective for the earth where the average surface temperature approaches 290 K.

1.3 Hydrostatic equation

Because a planet's atmosphere is in the planet's gravitational field, its density will fall with altitude. Since vertical motion is generally very small, the assumption of static equilibrium is a good starting point. If ρ is the density and p the pressure at altitude z measured vertically upwards from the surface we have

$$dp = -g\rho dz \qquad (1.2)$$

Since a planet's atmosphere is generally of total depth small compared with the planet's radius, the acceleration due to gravity g is approximately constant within the atmospheric region (cf. problem 1.5).

From the equation of state for a perfect gas of molecular weight M_r (see table 1.2) and temperature T,

$$\rho = \frac{M_r p}{RT} \tag{1.3}$$

where R is the gas constant per mole.

Equation (1.2), therefore, becomes

$$dp/p = -dz/H$$

which on integration gives, for the pressure p at altitude z,

$$p = p_0 \exp\left\{-\int_0^z dz/H\right\} \tag{1.4}$$

where p_0 is the pressure at $z = 0$, and $H = RT/M_r g$, known as the scale height, is the increase in altitude necessary to reduce the pressure by a factor e. For the lower atmosphere of the earth H varies between 6 km at $T = 210$ K to 8.5 km at $T = 290$ K. In the very high atmosphere, the molecular weight is no longer constant and the temperature is much higher, a situation which is discussed in more detail in §5.2.

1.4   Adiabatic lapse rate

The simplest model of a planetary atmosphere we can make is to assume that it is transparent to all radiation, that it contains no liquid particles and that the temperature of its lower boundary is that of the planet's surface whose mean temperature is determined by the simple calculation of §1.2. Consider the vertical motion of a 'parcel' at pressure p, temperature T and of specific volume V within such an atmosphere. We shall assume the atmosphere to be in hydrostatic equilibrium described by (1.2). Gravitational forces and buoyancy forces are, therefore, balanced and neither need be included explicitly in the expression for the first law of thermodynamics applied to unit mass which is therefore

$$dq = c_v\, dT + p\, dV \tag{1.5}$$

where c_v is the specific heat at constant volume. Providing no heat enters or leaves the parcel, the motion is adiabatic and the quantity of heat dq is zero.

Differentiation of the equation of state (1.3) (remembering that $1/V = \rho$) gives

$$p\,dV + V\,dp = R\,dT/M_r$$
$$= (c_p - c_v)dT \tag{1.6}$$

since for a perfect gas $c_p - c_v = R/M_r$ where c_p is the specific heat at constant pressure. On substituting for $p\,dV$ from (1.6) in (1.5) we have

$$c_p\,dT - V\,dp = dq = 0 \tag{1.7}$$

Fig. 1.2. (a) Illustrating the stability of a temperature profile *b*.
(b) Illustrating adiabatic lapse rate *a*; *inversion* condition *b*, i.e.
temperature increasing with height, an inversion occurs, for instance, when surface
has cooled more rapidly than the atmosphere; *superadiabatic* condition *c* which
cannot persist for very long and never occurs on a large scale.

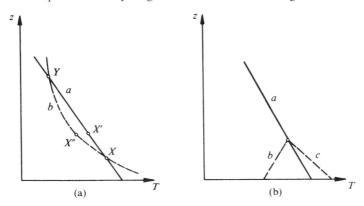

which gives, on substituting from (1.2),

$$\frac{dT}{dz} = -\frac{g}{c_p} = -\Gamma_d \tag{1.8}$$

Γ_d is known as the *adiabatic lapse rate* for a dry atmosphere. For the earth's atmosphere, $c_p = 1005\,\mathrm{J\,kg^{-1}\,K^{-1}}$ and $\Gamma_d \simeq 10\,\mathrm{K\,km^{-1}}$.

If the atmosphere, therefore, is heated by contact with the surface and vertical motion thereby ensues, we may expect a uniform temperature gradient with altitude of $10\,\mathrm{K\,km^{-1}}$.

It is informative to consider the stability with respect to vertical motion which occurs in an atmosphere with any temperature distribution *b* in fig. 1.2. If a parcel initially at *X* rises adiabatically it will follow the dry adiabatic *a* to *X'* where it will be surrounded by atmosphere under the conditions of *X''*, i.e. it will be warmer than the surroundings and will continue to rise. Point *X* on curve *b* is, therefore, unstable. By a similar argument point *Y* is stable. A given temperature gradient dT/dz represents stable or unstable conditions according as $-dT/dz$ is < or $> \Gamma_d$.

1.5   Sandström's theorem

Instead of assuming that the atmosphere is completely transparent to radiation, let us consider an atmosphere which absorbs strongly the incident solar energy being deposited in the upper atmosphere, as is likely, for instance, to be the case in the Venus atmosphere where a very deep

Fig. 1.3. Illustrating an adiabatic lapse rate above and an isothermal state below an absorbing layer.

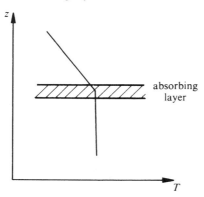

Fig. 1.4. A thermodynamic cycle releases energy when the source at temperature T_1 is at a higher pressure than the sink at temperature T_2.

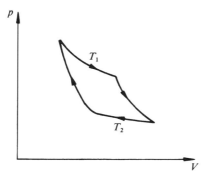

continuous cloud layer is present. What temperature structure is to be expected in the atmosphere below the absorbing layer? Because increasing temperature with altitude is a very stable situation (§1.4) vertical motion will not occur; conduction through the gas and radiation transfer will bring the lower atmosphere to an isothermal state (fig. 1.3).

This situation is an illustration of Sandström's theorem, which considers motion in the atmosphere to arise from the operation of a thermodynamic engine with heat sources and sinks. In a simple thermodynamic cycle (fig. 1.4), energy is released by going round the cycle in the direction of the arrows. This is only possible if the source is not only at a higher temperature than the sink but also, because for a gas at constant volume pressure increases with temperature, if the source is at a higher pressure than the sink. Hence Sandström's theorem which states that a closed steady circulation can only be maintained in an atmosphere if the heat source is situated at a higher pressure than the heat sink.

Because, in practice, heat is conducted from the precise place where it is deposited to the surrounding atmosphere and the sources and sinks to some extent are distributed by the motions themselves, strict interpretation of the theorem is difficult, as Jeffreys (1925) has pointed out. It is, however, certainly true that strong circulations cannot develop unless the conditions of Sandström's theorem are satisfied.

What then of the lower atmosphere of Venus which is observed to possess an adiabatic lapse rate of temperature from the cloud level to the surface which is at the surprisingly high temperature of ~750 K? We can only deduce that sufficient solar energy must penetrate to the surface for enough vertical motion to be developed to overcome the effects of conduction and radiation which would tend to make the profile isothermal – a suggestion which is, in fact, in agreement with measurements made by the Russian Venera probes of a small amount of sunlight being still incident on the probes when they reached the Venus surface.

Problems

(See the appendices for lists of constants and useful information about the atmosphere.)

1.1 If the planets of table 1.2 possessed no atmospheres, what would be the equilibrium temperatures of their surfaces, assuming a uniform albedo of the surfaces of 0.05?

1.2 From the information in table 1.2 compute the amount of energy per unit area arising from the internal source on Jupiter.

1.3 If the temperature falls uniformly with height z at $10 \, \text{K km}^{-1}$, write down an expression for the fall of pressure with height.

1.4 Calculate for the earth's atmosphere the height of the surface where the pressure is 0.1 of its surface value, assuming (1) uniform temperature of 290 K, (2) surface temperature of 290 K and uniform lapse rate of $10 \, \text{K km}^{-1}$.

1.5 What is the percentage change of the acceleration due to gravity g between the surface and 100 km altitude? Estimate the error in pressure determination at 100 km latitude through the use of (1.4) if a constant g is assumed.

1.6 The total mass of the oceans is 1.35×10^{21} kg. Compare this with total mass of the atmosphere. Also, compare the total heat capacity of the atmosphere with that of the oceans.

1.7 From the information in tables 1.1 and 1.2 compute the adiabatic lapse rates for Mars, Venus and Jupiter atmospheres.

1.8 For an unsaturated atmosphere containing 2% by volume of water vapour, find the value of the adiabatic lapse rate and compare it with the value of dry air.

1.9 A balloon filled with helium is required to carry a payload of 100 kg to a height of 30 km. What volume of balloon is required if the material used in its construction is polyethylene sheet, 25×10^{-4} cm thick and of density 1 g cm^{-3}?

1.10 Consider the vertical forces acting on a parcel displaced by a distance Δz from its equilibrium position in an atmosphere where the vertical temperature gradient is dT/dz. Write down the equation of motion in the form

$$\ddot{\Delta z} + gB\Delta z = 0$$

where $B = \dfrac{1}{T}\left(\dfrac{dT}{dz} + \Gamma_d\right)$

Discuss solutions to the equation of motion when $B < 0, = 0, > 0$. When $B > 0$, show that oscillations occur at the *Brunt–Vaisala frequency* $N_B = (gB)^{1/2}$. Calculate its value when $dT/dz = -6.5$ K km^{-1} – an average value for the lower part of the earth's atmosphere. This oscillation is a particular case of a *gravity wave*. These waves will be discussed further in §8.3.

2

A radiative equilibrium model

2.1 Black-body radiation

In fig. 2.1 are two curves showing the distribution with wavelength of the energy emitted by black-body sources at temperatures of 5750 K, approximating to that of the sun; and 245 K, a typical temperature for a planetary surface or a planetary atmosphere. Notice that the two curves are almost entirely separate. Also shown in fig. 2.1 is the absorption by various constituents in the earth's atmosphere; electronic bands occur in the ultraviolet and vibration–rotation and pure rotation bands due to minor constituents in the infrared. The major constituents nitrogen and oxygen, because of their symmetry, possess no electric dipole transitions and hence no strong bands due to them occur in the infrared.

In §1.4 an atmosphere was considered completely transparent to both solar and thermal radiation, in which heat was transferred from the heated surface by vertical convection. Inspection of fig. 2.1 suggests a slightly more elaborate model of an atmosphere transparent to solar radiation so that the surface is heated as before, but possessing an absorption coefficient uniform throughout the emitting infrared region and independent of pressure or temperature. Such an atmosphere is said to be *grey*.

The absorbing constituents in an atmosphere are molecules with complicated absorption spectra, much more complicated in fact than is indicated in fig. 2.1 (b) (cf. fig. 4.2). The approximation of a single absorption coefficient independent of wavelength, pressure and temperature is a very crude one. However, because of the large amount of overlap of the spectral lines involved, particularly in the lower atmosphere, the approximation is not, in fact, so crude as at first appears and it will enable us to derive features of the basic structure of an atmosphere in which radiative transfer is important. For a more accurate treatment and especially for

Fig. 2.1. (a) Curves of black-body energy B_λ at wavelength λ for 5750 K (approximating to the sun's temperature) and 245 K (approximating to the atmosphere's mean temperature). The curves have been drawn of equal areas since integrated over the earth's surface and all angles the solar and terrestrial fluxes are equal.

   (b) Absorption by atmospheric gases for a clear vertical column of atmosphere. The positions of the absorption bands of the main constituents are marked.

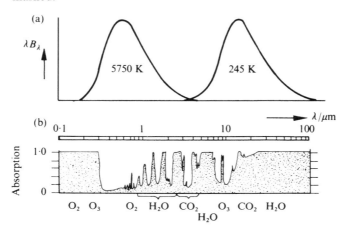

conditions in the upper atmosphere, more detailed consideration of the spectral structure is necessary. This will be given in chapter 4.

2.2   Absorption and emission

First it is necessary to derive some basic equations in radiative transfer. The equations will be derived for absorption and emission processes only. In the absence of clouds, scattering is unimportant at the infrared wavelengths we are considering and is, therefore, ignored for the moment. Radiative transfer in clouds is treated in chapter 6; in §6.4 it is shown that scattering processes may easily be incorporated into the same basic equations.

In a plane parallel atmosphere uniform in the horizontal consider first absorption of radiation along the vertical co-ordinate z. The law of absorption (sometimes known as Lambert's law or Bouguet's law) states that the absorption which occurs when radiation of intensity I (units: power per unit area per unit solid angle) traverses an elementary slab of atmosphere of thickness dz is proportional to the mass of absorber $\rho\,dz$ in unit cross-section of the slab (ρ being the density of absorber) and to the incident intensity of radiation itself. Thus

$$dI = -Ik\rho\,dz \tag{2.1}$$

where k is the absorption coefficient. Integrating (2.1) leads to

$$I = I_0 \exp(-\textstyle\int k\rho\,dz) \tag{2.2}$$

The quantity $\exp(-\int k\rho\,dz)$ is the *fractional transmission* τ of the path, and $\int k\rho\,dz$ is known as the *optical path* χ (when measured from the top of the atmosphere downwards it is the *optical depth*).

A slab of atmosphere will also emit radiation in amounts depending on its temperature. If thermodynamic equilibrium applies, application of Kirchhoff's law enables the amount emitted per unit area to be written as $k\rho\,dz\,B(T)$ where $B(T)$ is the black-body emission per unit solid angle per unit area of a surface at temperature T.

From the Stefan–Boltzmann law, the integral of $B(T)$ over a hemisphere is proportional to T^4, i.e.

$$\int_{2\pi} B\,dS\cos\theta\,d\omega = \sigma T^4 dS$$

giving $B = \pi^{-1}\sigma T^4$

where σ is the Stefan–Boltzmann constant and $d\omega$ is an element of solid angle at an angle θ to the normal to the element of surface area dS.

The earth's atmosphere is not, of course, precisely in thermodynamic equilibrium; however, in the lower atmosphere conditions known as *local thermodynamic equilibrium* (LTE) prevail which constitute a sufficiently good approximation to equilibrium for black-body emission to be employed in the equation for emission. We shall come back to this point later when considering the high atmosphere where LTE no longer applies (§5.7).

The equation for radiative transfer through the slab, which includes both absorption and emission, is sometimes known as *Schwarzschild's equation*:

$$dI = -Ik\rho\,dz + Bk\rho\,dz$$

or $\quad \dfrac{dI}{d\chi} = I - B \tag{2.3}$

In deriving (2.3), vertically travelling radiation only has been considered. Radiation is, of course, travelling in all directions. However, under the assumption of a plane parallel atmosphere the problem is reduced to a one dimensional one by considering the two fluxes $F^\uparrow$ and $F^\downarrow$ which are the quantities $\int I(\theta)\cos\theta\,d\omega$ integrated over the downward facing and upward facing hemispheres respectively, $I(\theta)$ being the intensity at an angle θ to the vertical and $d\omega$ an element of solid angle (fig. 2.2). Detailed

Fig. 2.2

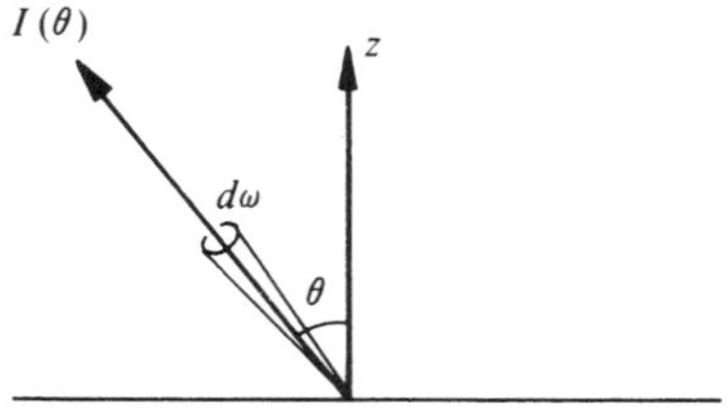

Fig. 2.3

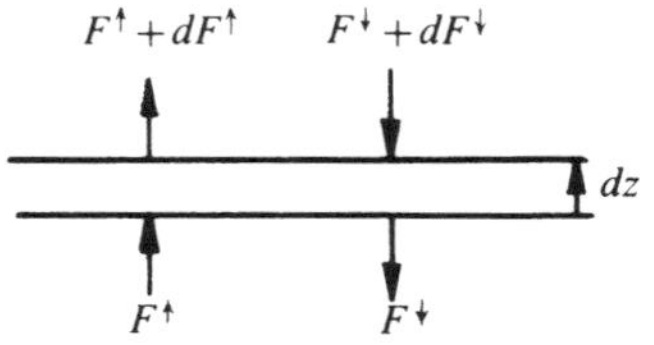

calculation shows that to a good approximation in (2.3) I may be replaced by F if dz is replaced by a mean thickness $\frac{5}{3}dz$ and B is replaced by πB, the black-body function integrated over a hemisphere.

2.3   Radiative equilibrium in a grey atmosphere

Having now laid the theoretical foundation of radiative transfer in a grey atmosphere, we apply it to find the equilibrium situation in an atmosphere in which transfer of infrared radiation is the only energy transfer mechanism and which has for its lower boundary a heated surface at temperature T_g.

Under these assumptions the net rate of temperature change, dT/dt, which occurs in a slab of plane parallel atmosphere due to the divergence of the upward and downward radiation fluxes is given by (fig. 2.3)

$$\frac{d}{dz}\left(F^{\downarrow} - F^{\uparrow}\right) = \rho c_p \frac{dT}{dt} \tag{2.4}$$

In equilibrium $dT/dt = 0$, and integration of (2.4) gives

$F^{\uparrow} - F^{\downarrow} = $ a constant ϕ, the net flux. $\tag{2.5}$

Further, from (2.3) the transfer equations are

$$\frac{dF^{\uparrow}}{d\chi^*} = F^{\uparrow} - \pi B$$

$$-\frac{dF^{\downarrow}}{d\chi^*} = F^{\downarrow} - \pi B \tag{2.6}$$

where the star on the optical depth χ^*, which is measured from the top of the atmosphere, denotes that it is appropriate to hemispheric radiation (i.e. the factor $\frac{5}{3}$ has been applied to dz in the expression for χ).

If

$$\psi = F^\uparrow + F^\downarrow \tag{2.7}$$

equations (2.6) may be written

$$\frac{d\psi}{d\chi^*} = \phi \tag{2.8}$$

$$\frac{d\phi}{d\chi^*} = \psi - 2\pi B \tag{2.9}$$

Since $\phi = $ constant (from (2.5)) $d\phi/d\chi^* = 0$

and $\quad \psi = 2\pi B \tag{2.10}$

which on substitution into (2.8) gives

$$B = \frac{\phi}{2\pi} \chi^* + \text{constant} \tag{2.11}$$

The boundary condition at the top of the atmosphere ($\chi^* = 0$) is $F^\downarrow = 0$, so that here $\psi = \phi$ and, from (2.10), the constant in (2.11) is $\phi/2\pi$, i.e.

$$B = \frac{\phi}{2\pi}(\chi^* + 1) \tag{2.12}$$

At the bottom of the atmosphere where $\chi^* = \chi_0^*$, $F^\uparrow = \pi B_g$, B_g being the black-body function at the temperature of the ground. It is easy to show that there must be a temperature discontinuity at the lower boundary, the black-body function for the air close to the ground being B_0, and

$$B_g - B_0 = \frac{\phi}{2\pi} \tag{2.13}$$

The black-body function B for such an atmosphere is plotted against χ^* in fig. 2.4. The equivalent diagram in terms of temperature and height is fig. 2.5. Notice that the lower part of the atmosphere possesses a very steep lapse rate of temperature, and at the surface itself there is a discontinuity in temperature. As was shown in §1.4, such a steep lapse rate is very unstable with respect to vertical motion, and will soon be destroyed by the process of *convection* which will tend to establish a mean adiabatic lapse rate.

In fig. 2.5 air from near the surface will tend to rise along the line which has been drawn with a slope of $-6\,\mathrm{K\,km^{-1}}$ (a mean adiabatic lapse rate, see §1.4 and also §3.2) and which cuts the radiative equilibrium curve at the height of approximately 10 km.

Fig. 2.4. Upward radiation flux $F^\uparrow$, downward flux $F^\downarrow$ and black-body function πB at atmospheric temperature plotted against optical depth for radiative equilibrium atmosphere.

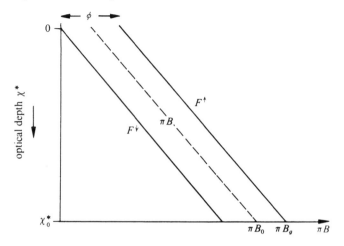

Fig. 2.5. Radiative equilibrium temperature T plotted against altitude z. The line c is drawn through the surface temperature with slope $-6\,\mathrm{K\,km^{-1}}$. This simple model leads to a troposphere dominated by convection below a stratosphere in approximate radiative equilibrium.

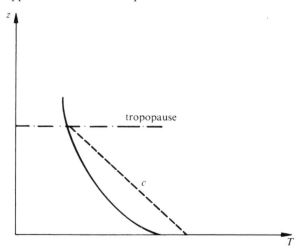

This simple model does, in fact, bear some relation to what occurs in the real atmosphere. The *tropopause* is a surface situated at a height of ~10 km in mid latitudes which divides the region below (the *troposphere* or turning-sphere), in which convection is the dominant mechanism of vertical heat transfer, from the *stratosphere* which is much more stably stratified and where radiative transfer is dominant.

2.4   Radiative time constants

In considering radiative processes it is important to know the rate at which changes may be expected to occur. The time constant appropriate to any particular process will depend on the scale of the phenomenon under consideration, and the density of the air at the altitude in question, as well as on the details of the process itself. For the atmosphere as a whole a rough estimate may be made by considering a slab of atmosphere of thickness h and uniform density ρ radiating like a black body at a temperature $T \simeq 270\,K$, different by a small amount ΔT from what it would be if in radiative equilibrium with the surrounding layers both above and below the slab.

Its rate of change of temperature is given by

$$c_p \rho h \frac{d\,\Delta T}{dt} = 8\sigma T^3 \Delta T \tag{2.14}$$

The time constant resulting from (2.14) is $c_p \rho h / 8\sigma T^3$. If ρ is the density appropriate to the 50 kPa level and h equal to the scale height H, i.e. ~8 km, the magnitude of this time constant is about 6 days. Radiative processes in the lower atmosphere, therefore, in general, act rather slowly and for the consideration of short-term atmospheric developments may often be neglected. In the longer term, however, because they are the processes which largely determine the distribution of energy sources and sinks they are of dominant importance.

2.5   The greenhouse effect

Combining (2.12) and (2.13) we find that for the radiative equilibrium atmosphere:

$$B_g = \frac{\phi}{2\pi}\left(\chi_0^* + 2\right) \tag{2.15}$$

where χ_0^* is the optical depth at the bottom of the atmosphere. If $\chi_0^* = 0$, $B_g = \phi/\pi$ and the surface temperature is in equilibrium with the incoming and the outgoing radiation, which are both equal to ϕ. If χ_0^* is large, the surface temperature represented by the black-body function B_g will be very considerably enhanced, an illustration of the *greenhouse effect* mentioned in §1.2. It will be considered in more detail in chapter 14.

In practice, the optical depth of an atmosphere will depend on the concentration of gases, such as water vapour, which are evaporated from

Fig. 2.6. Illustrating the greenhouse effect for the terrestrial planets. Their surface temperature is plotted against the vapour pressure of water vapour in the atmosphere. Also on the diagram (dashed) are the phase lines for water, the shaded area showing where liquid water is in equilibrium. For Mars and the earth the greenhouse effect is halted when water vapour becomes saturated with respect to ice or water. For Venus the diagram illustrates the runaway greenhouse effect. (After Rasool & De Bergh, 1970, and Goody & Walker, 1972)

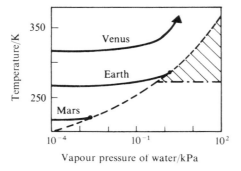

the surface and whose concentration may increase rapidly with surface temperature. A mechanism for positive feedback therefore exists which may be particularly important in the case of the Venus atmosphere (Rasool & DeBergh, 1970; Ingersoll, 1969). It has been called the *runaway greenhouse effect* and is illustrated for three planets in fig. 2.6. Suppose the atmospheres of these planets began to form by outgassing from their interiors at a time when their surface temperatures were essentially determined by equilibrium between absorbed solar radiation and emitted long-wave radiation ((1.1) with $A \simeq 0$ because no clouds would be present) at values given by those on the left hand side of fig. 2.6. As water vapour accumulated in an atmosphere, owing to the greenhouse effect the surface temperature rose, so increasing the evaporation from the surface until either the atmosphere became saturated with water vapour or all the available water had evaporated. The formation of clouds by condensation of water vapour enhances the effect except for the case of Mars where the atmosphere is so thin that significant cloud formation has not occurred and the blanketing effect of the atmosphere is small. For the earth, the equilibrium situation is with most of the water in liquid form, while for Venus, on these assumptions, the surface temperature would always be above the boiling point of water at the surface pressure. No liquid water would, therefore, be present on Venus. Supposing that originally similar amounts of water were present as for the earth, the early Venus atmosphere would have water vapour as the major constituent. No other gases would be present to prevent ultraviolet solar radiation dissociating water vapour at the top of the atmosphere; the resulting hydrogen would escape (§5.3) and the oxygen would be consumed in various oxidation processes at the surface. An explanation is, therefore, afforded of the small quantity of water on Venus today compared with the large amount on the earth. The large amount of carbon dioxide remaining in the Venus atmosphere rather than as carbonates in the rocks is also consistent with this atmospheric history.

Problems

2.1 If one third of the solar energy incident outside the atmosphere were absorbed by the atmosphere, show that the average rate of temperature rise (assuming no loss) of the atmosphere would be ~1 K per day. Rates of temperature change due to radiative processes are often expressed in units of K per day. For instance an average rate of loss of energy by the troposphere due to emission of long-wave radiation is ~1 K per day.

2.2 Assuming an absorber with uniform absorption coefficient and uniformly mixed with height, a surface temperature of 280 K, a tropopause temperature at 25 kPa pressure of 220 K and radiative equilibrium conditions, find ϕ. Also find χ^* in equation (2.12) as a function of pressure. What would be the temperature discontinuity at the surface under such assumptions?

2.3 The calculation in §2.3 has been carried out on the assumption that radiative equilibrium exists throughout the atmosphere. Make a qualitative assessment of the effect on the radiative equilibrium profile above the tropopause of a change in the profile below the tropopause from the radiative equilibrium one to the convective one (fig. 2.5).

2.4 Estimate the radiative time constant for the atmosphere of Mars.

2.5 What is the length of a solar day on Venus? (cf. table 1.2 noting also that the sense of rotation of Venus is opposite to that of the earth and Mars). Ignoring the effect of the clouds, by a similar calculation to problem 2.4, estimate the radiative time constant for the lower Venus atmosphere. Because of the blanketing effect of the clouds (fig. 2.7) and probably also because of motion in the Venus atmosphere, no difference in temperature in the lower

Fig. 2.7. An idea of the variability of cloud features on Venus can be obtained from these four images taken in the ultraviolet from the Pioneer Venus orbiter over a period of 38 hours in May 1980. Note the polar rings which are brighter than the other cloud features because of the greater number at these latitudes of small particles which scatter ultraviolet light more effectively (cf. §4.1). (From Briggs & Taylor, 1982)

Venus atmosphere has been measured between the day and the night side.

2.6   The surface temperature of Venus is ~750 K (table 1.1). From the value of T_m from table 1.2 and from (2.15) assuming radiative equilibrium, calculate the optical depth of the Venus atmosphere.

3

Thermodynamics

3.1 Entropy of dry air

In chapter 1 the first law of thermodynamics was applied to vertical motions in a planetary atmosphere and the adiabatic lapse rate for dry air was derived. On substituting TdS for dq in (1.7), where S is the entropy, and substituting for V from the equation of state for a perfect gas we have

$$dS = c_p \frac{dT}{T} - \frac{R\,dp}{M_r p} \qquad (3.1)$$

which on integration gives an expression for the entropy per unit mass:

$$S = c_p \ln T - RM_r^{-1} \ln p + \text{constant} \qquad (3.2)$$

If air at pressure p and temperature T is brought adiabatically (i.e. at constant S) to a standard pressure p_0 of 100 kPa, its temperature θ at that standard pressure is known as its *potential temperature*.

From (3.2)

$$c_p \ln \theta = c_p \ln T - RM_r^{-1} \ln p + RM_r^{-1} \ln p_0 \qquad (3.3)$$

from which, remembering that for a perfect gas $c_p - c_v = R/M_r$,

$$\theta = T \left(\frac{p_0}{p} \right)^\kappa \qquad (3.4)$$

where $(c_p - c_v)/c_p = \kappa = 0.288$ for dry air in the earth's atmosphere.

Also

$$S = c_p \ln \theta + \text{a constant} \qquad (3.5)$$

– a useful equation because it enables us to express the entropy of dry air in terms of the more readily interpretable concept of potential temperature.

3.2   Vertical motion of saturated air

The structure of an atmosphere may be influenced very considerably by the presence within it of a condensable vapour. In the case of the
earth's atmosphere water vapour is very important.

Consider the vertical motion of air containing water vapour. Provided air remains unsaturated, its thermodynamic properties are little
different from those of dry air (problem 3.1). As unsaturated air rises,
therefore, it cools similarly to dry air at about $10\,\mathrm{K\,km^{-1}}$. During this
ascent although the *mixing ratio*[†] of water vapour remains constant, the
relative humidity[‡] increases and may reach 100%. The level at which this
occurs is the *condensation level*. As the air continues to rise, it remains
saturated, the surplus water vapour condensing to form drops of liquid
water. The latent heat released by this condensation process must now
be included.

The same assumptions are made as in §1.4, together with the additional assumption that the parcel contains liquid water which moves with
it; the air remains saturated, evaporation and condensation occurring so
that equilibrium is continually maintained. Consider a parcel of 1 g of dry
air with m g of water vapour (m is the *saturation mixing ratio*) and $\xi - m$
g of liquid water. For the dry air the entropy is given by (3.2). The entropy
of the water vapour and liquid water is the entropy of ξ g of liquid water
at temperature T plus the additional entropy Lm/T required to convert m
g of liquid water to water vapour. The total entropy S is therefore

$$S = (c_p + \xi c)\ln T - \frac{R}{M_{ra}}\ln(p - e) + \frac{Lm}{T} + \text{constant} \qquad (3.6)$$

where c is the specific heat of liquid water, e the saturation vapour pressure and L the latent heat which, of course, varies with temperature. Differentiating (3.6) and substituting from the hydrostatic equation which for
wet air is

$$dp = -g\rho_a(1 + \xi)dz \qquad (3.7)$$

we have, for adiabatic conditions,

$$(c_p + \xi c)\frac{dT}{T} + d\left(\frac{Lm}{T}\right) + \frac{R}{M_{ra}(p - e)}\frac{de}{dT}dT + \frac{g}{T}(1 + \xi)dz = 0 \qquad (3.8)$$

[†] i.e. the ratio of number of water molecules to other molecules in a given volume (the
volume mixing ratio) or the ratio of the mass of water vapour to mass of air in a given
volume (the mass mixing ratio).

[‡] the ratio of water vapour pressure to saturation water vapour pressure at air
temperature.

The equations of state for unit masses of the air (density ρ_a, molecular weight M_{ra}) and water vapour (density ρ_v, molecular weight M_{rv}) considered separately are

$$p - e = \rho_a RT/M_{ra} \qquad (3.9)$$

and $\quad e = \rho_v RT/M_{rv} \qquad (3.10)$

Now the mixing ratio m is dependent on air pressure as well as temperature

$$m = \frac{\rho_v}{\rho_a} = \left(\frac{e}{p-e}\right)\frac{M_{rv}}{M_{ra}} \simeq \frac{e\varepsilon}{p} \qquad (3.11)$$

since, in the lower atmosphere, $e \ll p$, and

$$\varepsilon = M_{rv}/M_{ra} = 0.622$$

Differentiating (3.11) gives

$$\frac{dm}{dz} = \frac{\varepsilon}{p}\frac{de}{dT}\cdot\frac{dT}{dz} - \frac{\varepsilon e}{p^2}\cdot\frac{dp}{dz} \qquad (3.12)$$

Substituting from (3.12) in (3.8), considering the air just saturated, i.e. $\xi = m$, and substituting from the Clausius Clapeyron equation

$$\frac{de}{dT} = \frac{LeM_{rv}}{RT^2} \qquad (3.13)$$

we have

$$-\frac{dT}{dz} = \Gamma_s$$

$$= \frac{g}{c_p}\cdot\frac{\{1+(LeM_{rv}/pRT)\}\{1+(e\varepsilon/p)\}}{[1+(\varepsilon e/pc_p)\{c+(dL/dT)\}+(\varepsilon eL^2 M_{rv}/c_p pRT^2)]} \qquad (3.14)$$

On inserting typical values (problem 3.4) the middle term in the denominator on the right hand side turns out to be negligible in comparison with the other terms. Also, since $e \ll p$, the second bracket in the numerator $\simeq 1$, so that (3.14) becomes

$$\Gamma_s = \Gamma_d\left(1+\frac{LeM_{rv}}{pRT}\right)\left(1+\frac{LeM_{rv}}{pRT}\cdot\frac{\varepsilon L}{c_p T}\right)^{-1} \qquad (3.15)$$

Γ_s is always $<\Gamma_d$ and varies in the atmosphere from about $0.3\Gamma_d$ to Γ_d. Actual values can be calculated from the formula in problem 3.3 or read from fig. 3.1. For a shorter derivation of (3.15), see problem 3.2.

The stability of saturated air containing liquid water for vertical motion can be considered in exactly the same way as that for dry air in

§1.4. We have stable or unstable conditions according as $-dT/dz$ is $<$ or $> \Gamma_s$. It should be noted that for saturated air with no liquid water content the stability condition for upward motion involves Γ_s, while that for downward motion involves Γ_d, since as soon as the air moves downward it becomes unsaturated.

The assumption made above, that the liquid water remains with the parcel as it is lifted, is clearly not necessarily fulfilled in practice. If, however, it is assumed that the products of condensation, or a proportion of them, fall out during the ascent, little difference is made to the final answer. Another assumption which has been made in considering vertical motion is that the environment is unmodified by the rising or sinking parcels of air. The motion of a parcel of air is, of course, bound to produce a compensating motion in the environment. However, since the compensating descending motions normally occur over a much larger area than the ascending motion, the *parcel* method we have described, which ignores the compensating motions, is in general a good approximation and is used in general forecasting analysis.

3.3   The tephigram

The tephigram is a thermodynamic diagram particularly suitable for representing atmospheric processes. The name derives from $T–\phi$-gram since it is a temperature–entropy plot, the symbol ϕ sometimes being used for entropy (we have used the symbol S). Instead of plotting S as the ordinate in our thermodynamic diagram, we can plot $\ln \theta$ (3.5) remembering that θ is the potential temperature referred to *dry* air. The basic diagram therefore is a plot of $\ln \theta$ against T. Contours of other quantities may also be plotted on the tephigram to assist in its use, namely:

(1) The pressure p, from (3.4), is a function of T and θ, so isobars or lines of constant pressure can be drawn. For many applications it is then convenient to use the diagram as one in which T can be plotted as a function of p.

(2) Adiabatic lines for dry air are lines of constant θ.

(3) Adiabatic lines for saturated air may be plotted from (3.15) where the lapse rate is given as a function of T and p. These *wet adiabatics* show the entropy change of the saturated air which is exactly equal but opposite in sign to that of the liquid water carried with it (or falling out of it).

(4) The saturation mixing ratio m from (3.11) is a function of T and p only. Note that a plot of mixing ratio against pressure for unsatu-

Fig. 3.1. Part of a tephigram. (Charts available in the UK from Her Majesty's Stationery Office.)

The main axes of entropy and temperature are inclined. Isopleths of pressure (marked in mb, 1000 mb = 100 kPa) are then nearly parallel to the lower edge of the page.

The dashed lines are lines of constant water vapour mixing ratio (labelled in g kg^{-1}) and the most curved lines are wet adiabatics.

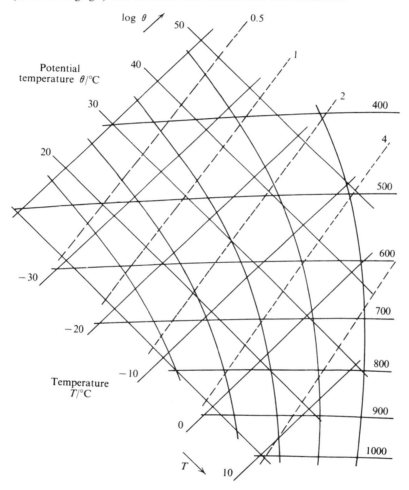

rated air, is identical to a plot of *dew point* (or frost point for temperatures below zero celsius) against pressure).

Fig. 3.1 is an example of a tephigram on which the above isopleths are drawn. Problems at the end of the chapter illustrate its use in some detail. Also figs. 3.4, 3.5 and 3.6 illustrate from satellite and radar pictures some of the thermodynamic processes involved in the atmosphere.

3.4   Total potential energy of an air column

The energy dE of an element of a static air column at altitude z consists of two parts, its internal energy dE_I which is $c_v T$ per unit mass and its potential energy dE_P which is gz per unit mass. Since an element of mass is $-dp/g$ (1.2) we have

$$E = E_I + E_P = g^{-1}\int_{p_h}^{p_0} c_v T \, dp + \int_{p_h}^{p_0} z \, dp \qquad (3.16)$$

where the column extends from $z = 0$, $p = p_0$ to $z = h$, $p = p_h$. The second term on the right hand side may be integrated by parts to give

$$E_P = -p_h h + \int_0^h p \, dz \qquad (3.17)$$

Now since $p = R_\rho T/M_r$ (equation of state (1.3))

and $\rho dz = -dp/g$, we have

$$E_P = -p_h h + g^{-1}\int_{p_h}^{p_0} \frac{RT}{M_r} dp \qquad (3.18)$$

The first term may be made negligible by choosing h sufficiently high that $p_h \simeq 0$. Substituting from (3.18) into (3.16), remembering that $c_p = c_v + R/M_r$, we have

$$E = E_I + E_P = c_p g^{-1}\int_0^{p_0} T \, dp \qquad (3.19)$$

Note that $E_P/E_I = c_p/c_v - 1 \simeq 0.4$ for dry air so that, to the extent that hydrostatic equilibrium prevails, the potential and internal energies of a column of air bear a constant ratio to each other. Conversion into kinetic energy will, therefore, occur at the expense of both potential and internal energy, in this same ratio. It is, therefore, convenient to treat potential and internal energy together; their sum is called *total potential energy* – a name first used by Margules in a famous paper in 1903 on the energy of storms.

3.5   Available potential energy

All the total potential energy is clearly not available for conversion into kinetic energy. For instance, for a uniformly stratified atmosphere, stable with respect to vertical motion, in which there are no variations of density in the horizontal at any level, although the total potential energy is large, none at all is available for conversion. Suppose now that such a stratified atmosphere is heated in a restricted region. Total potential energy is added and the stratification is disturbed; horizontal density gradients and hence pressure forces are created which may convert total potential energy into kinetic energy. Suppose, further, that a uniformly stratified atmosphere is cooled over a limited region rather than heated. Although total potential energy is removed, as a result of this removal the stratification is disturbed and conversion into kinetic energy is again possible. Removal of energy can be as effective as addition of energy in making more energy available.

A quantity which for any given atmospheric state is a measure of the total potential energy which is available for conversion through adiabatic processes into kinetic energy was introduced by Lorenz (1955) and is called *available potential energy*. It is the difference between the total potential energy in the state under consideration, and the total potential energy of a state of uniform stratification which is statically stable (i.e. having a stable lapse rate of temperature) obtained by a redistribution of atmospheric mass by adiabatic processes. For conditions of such uniform stratification the available potential energy is zero.

The process of readjustment to the statically stable reference state involves the imposition of vertical motions so that the originally undulating isentropic surfaces (i.e. surfaces of constant potential temperature) are brought into coincidence with surfaces of constant geopotential (cf. §7.5 and problem 7.11 for definition of geopotential). Since warmer air rises and cooler air descends during the readjustment, gravitational potential energy will be released and work will be done at the expense of internal energy; the sum of these is the available potential energy we wish to find.

To calculate its magnitude, following Lorenz (1955), consider the atmosphere divided up by surfaces of constant potential temperature θ (§3.1), such as those illustrated by the example of fig. 3.2.

In any redistribution under adiabatic flow the value of θ for any given element of atmosphere will be conserved. Provided that each of the surfaces of constant θ intersect every vertical column only once, we can easily define the average pressure $\bar{p}$ over each of the surfaces with respect to the area projected on to a horizontal surface. Surfaces which appear to intersect the ground may be imagined to continue at ground level, i.e. where $p = p_0$. The total potential energy E of a vertical column (3.19) can be written in terms of potential temperature θ from (3.4) and integrated by parts to give

$$E = (1+\kappa)^{-1} c_p g^{-1} p_0^{-\kappa} \left(\int_{\theta_0}^{\infty} p^{1+\kappa} d\theta + \theta_0 p_0^{1+\kappa} \right) \tag{3.20}$$

Provided that the value of θ at the surface, namely θ_0, is lower than any value in the atmosphere above (this is a requirement for stability; see also fig. 3.2), the last term on the right hand side of (3.20) can be incorporated

Fig. 3.2. (a) Average temperature cross-section of atmosphere for northern hemisphere winter.

   (b) Distribution of potential temperature θ corresponding to temperature distribution in (a).

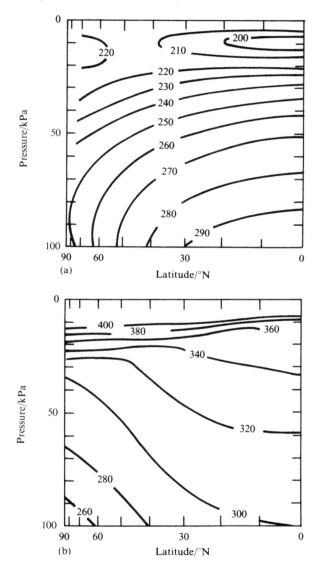

Fig. 3.2. (*continued*) (c) Potential temperature distribution which results when atmosphere shown in (b) is moved isentropically to a state of uniform horizontal stratification. (After Lorenz, 1967)

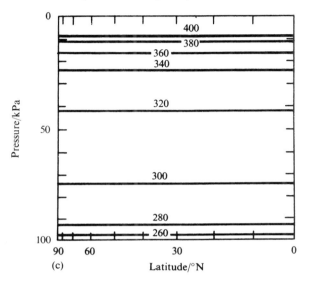

in the integral by replacing the lower limit with $\theta = 0$ with the interpretation that all surfaces of θ with values less than θ_0 coincide at $p = p_0$. We therefore have for the whole atmosphere (per unit area of surface)

$$E = (1+\kappa)^{-1} c_p g^{-1} p_0^{-\kappa} \int_0^\infty \overline{p^{1+\kappa}} d\theta \qquad (3.21)$$

where $\overline{p^{1+\kappa}}$ is the average of $p^{1+\kappa}$ over an isentropic surface, the average to be taken in the same way as when defining the average pressure $\bar{p}$.

The minimum total potential energy which can result from adiabatic rearrangement occurs when $p = \bar{p}$ everywhere over a given isentropic surface. The difference between this minimum value of total potential energy and the original value is the *available potential energy* E_A and is, therefore, given by

$$E_A = (1+\kappa)^{-1} c_p g^{-1} p_0^{-\kappa} \int_0^\infty \left(\overline{p^{1+\kappa}} - \bar{p}^{1+\kappa} \right) d\theta \qquad (3.22)$$

Because $1 + \kappa > 1$, $\overline{p^{1+\kappa}} > \bar{p}^{1+\kappa}$ unless $p \equiv \bar{p}$, so that (3.22) ensures that E_A is positive, as is, of course, necessary.

In (3.22) let $p = \bar{p} + p'$, $p^{1+\kappa}$ can then be expanded by the binomial theorem. Remembering that $\bar{p'} = 0$ and expanding to the first term only,

$$E_A = \frac{1}{2} \kappa c_p g^{-1} p_0^{-\kappa} \int_0^\infty \overline{p}^{1+\kappa} \overline{\left(\frac{p'}{\overline{p}}\right)^2} \, d\theta \tag{3.23}$$

Equation (3.23) expresses E_A in terms of the variance of p on an isentropic surface. It is often more convenient to refer to isobaric surfaces, in which case an appropriate expression for E_A is (problem 3.12)

$$E_A = \frac{1}{2} c_p g^{-1} \int_0^\infty \overline{T} \left(1 - \frac{\Gamma}{\Gamma_d}\right)^{-1} \overline{\left(\frac{T'}{\overline{T}}\right)^2} \, dp \tag{3.24}$$

Equation (3.24) is a useful form for estimating E_A. By comparing it, for instance, with (3.19), an estimate can be made of the ratio of available potential energy E_A to total potential energy E. Typical values in (3.24) are $\Gamma \simeq \frac{2}{3}\Gamma_d$ and $\overline{(T'/\overline{T})^2} \simeq (\frac{1}{16})^2$ giving $E_A/E \simeq \frac{1}{200}$. Only about $\frac{1}{2}\%$ of the total potential energy, therefore, is available for conversion into kinetic energy.

It is interesting to compare the average kinetic energy with both the total potential energy and the available potential energy. The total potential energy of a vertical column (3.19) may be written as

$$E = \{(\gamma - 1)g\}^{-1} \int_0^{P_0} c^2 dp \tag{3.25}$$

where $\gamma = c_p/c_v$ and $c = (\gamma R T)^{1/2}$ is the speed of sound in air. The kinetic energy E_K of a vertical column expressed in a similar form is

$$E_\kappa = \frac{1}{2g} \int_0^{P_0} V^2 dp \tag{3.26}$$

where V is the velocity of atmospheric motion. A typical value of V is ~$0.05c$ so that the ratio E/E_K may be ~2000 on average.

Since $E/E_A \simeq 200$, $E_K/E_A \simeq 0.1$ on average, so that the amount of available potential energy in the atmosphere is much larger than the amount of kinetic energy.

The concept of available potential energy may be employed to investigate the efficiency of heating or cooling different parts of the atmosphere in contributing to the maintenance of its general circulation. If the quantity $\dot{q}\,dm$ (dm being an element of mass) represents the rate at which heat is introduced or removed by diabatic processes (e.g. radiation, latent heat, turbulent transfer from the surface etc.), it can be shown that (see, for instance, Dutton & Johnson, 1967) the rate of generation of available potential energy is approximately

$$\int \left\{1 - \left(\frac{\overline{p}}{p}\right)^\kappa\right\} \dot{q} \, dm \tag{3.27}$$

Lorenz called the quantity $\{1 - (\bar{p}/p)^\kappa\}$ the efficiency factor. Inspection of fig. 3.2 (problem 3.14) shows that maximum generation of available potential energy will occur with low altitude heating in low latitudes and high altitude cooling in polar regions – a result which might be expected from the very general thermodynamic arguments of §1.5.

In the above discussion the available potential energy has been computed by reference to a *statically stable* reference state. Since there are dynamical constraints on the motions which may occur in the atmosphere (see chapters 7ff.) it is not clear that processes for attaining such a reference state are dynamically possible. In fact, as has been suggested by Van Mieghem (see Dutton & Johnson, 1967), it may be more realistic to postulate a reference state which includes a circumpolar vortex balanced by a horizontal temperature gradient which may be reached by isentropic flows in the manner described above, but which also preserves the absolute zonal angular momentum.

3.6   Zonal and eddy energy

The above discussion has not been concerned at all with the way in which the conversion from potential energy to kinetic energy takes place. In chapter 10, two mechanisms for such conversion will be described, namely (1) through mean motions which may be considered independent of longitude, and (2) through large scale quasi-horizontal eddy motions which show large variations with longitude. For studies of the general circulation, therefore, it is convenient to divide the kinetic energy into two components by resolving atmospheric motion into (1) the mean zonal component, i.e. the wind velocity averaged around latitude circles, and (2) the eddy components, i.e. the meridional (or north–south) component and the deviation of the zonal component at any place from the mean around the latitude circle. These components are known as the zonal kinetic energy and eddy kinetic energy respectively.

In a similar way zonal and eddy available potential energy may be defined. Components of the variance of the temperature field are chosen, namely (1) the variance of the zonally averaged temperature, and (2) the variance of temperature along latitude circles, from which zonal and eddy components of available potential energy may be deduced.

Fig. 3.3 shows an example of the result of a study of energy generation and conversion. Notice that most of the conversion to kinetic energy occurs from the eddy available potential energy which in turn is converted through large scale eddy motions from zonal available potential energy.

Fig. 3.3. Average storage and conversions of available potential energy E_A and kinetic energy E_K for the whole atmosphere as estimated by Oort & Peixoto (1974). Units of energy in $10^5\,\mathrm{J\,m^{-2}}$ and of conversion, generation or dissipation in $\mathrm{W\,m^{-2}}$. Subscripts Z and E are for zonal and eddy components respectively.

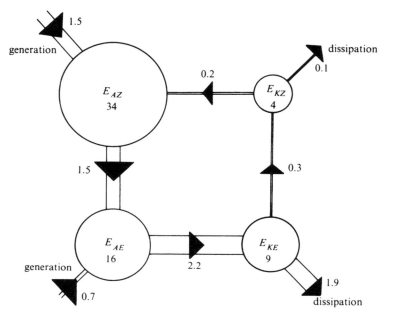

Problems

3.1 To allow in a convenient way for the different molecular weight M_{rw} which applies to air containing water vapour, a *virtual temperature* T^* is defined such that

$$RT^*/M_{ra} = RT/M_{rw}$$

where M_{ra} is the molecular weight of dry air. Derive an expression for the virtual temperature in terms of water vapour mixing ratio and find $T^* - T$ for saturated air at 100 kPa and 273 K, 290 K and 300 K.

3.2 Derive the expression (3.15) for the saturated adiabatic lapse rate starting from (1.7) with $dq = -L\,dm$ (this neglects any variation of L with T which in this context is small). Then substitute for dm from (3.12) and obtain (3.15) for Γ_s.

3.3 An approximate empirical formula for Γ_s in $\mathrm{K\,km^{-1}}$ is (Gill 1982)

$$\Gamma_s = 6.4 - 0.12t + 2.5 \times 10^{-5}t^3 + \left[10^{-3}(t-5)^2 - 2.4\right][1 - 0.001p]$$

where t is the temperature in °C and p the pressure in *mb*.

From this formula, calculate values of Γ_s for $p = 1000\,mb$, $t = 30°C$, $20°C$ and $10°C$ and for $p = 500\,mb$, $t = -20°C$.

3.4 Calculate typical values of the middle term in the denominator of (3.14) and show that for all atmospheric conditions, it is $\ll 1$. [dL/dT is approximately $-2\,J\,g^{-1}\,K^{-1}$.]

3.5 Calculate values of the term $\varepsilon L/c_p T$ in (3.15) and show that because it is always >1, $\Gamma_s < \Gamma_d$.

3.6 From the hydrostatic equation (1.2) show that the pressure $p(z)$ at height z is related to the field of potential temperature $\theta(z')$ by the equation

$$\left\{\frac{p(z)}{p(0)}\right\}^{\kappa} = \frac{\kappa g M_r}{R} \int_z^{\infty} \frac{dz'}{\theta(z')}$$

where $\kappa = (\gamma - 1)/\gamma$.

3.7 The following measurements were made at 0001 GMT on a day in June by a radiosonde ascent from Liverpool:

Pressure (kPa)	Temperature (°C)	Dew (or frost) point (°C)
100	13	11
94	9.5	8
90	7	5
78	0	-3
70	-5	-11
60	-11	-17
50	-20	-28
40	-32	-42
30	-47	
20	-49	
15	-50	
10	-48	

Plot these on a tephigram chart and answer the following questions:

(1) What is the pressure at the tropopause?
(2) What parts of the ascent are stable for (a) dry air, (b) saturated air?
(3) What is the mixing ratio, water vapour : air, at $100\,kPa$, $50\,kPa$?
(4) If the surface cools radiatively during the night, how many degrees of cooling are required for fog to begin to form?
(5) If air at the surface is heated during the following day and then rises adiabatically, at what level will condensation occur?

(6)   What will be the level of the top of convective clouds which develop during the day?

3.8 Air represented by the measurements given in problem 3.7 is forced to ascend either over a range of mountains or over a wedge of colder air at a cold front. Consider the layer of air initially between 78 and 70 kPa; note that it is stable for both dry and wet ascent. Suppose this air is forced to rise by 10 kPa in pressure. Note that it will remain 8 kPa deep in pressure units. Using a tephigram chart find the new temperature of the bottom of the ascent by taking it up the dry adiabatic until it becomes saturated, then up the wet adiabatic. Do the same for the top of the ascent. Hence show that the layer is now unstable with respect to wet ascent. Such a situation is called one of *potential instability*; the amount of such instability will determine how much convective development will occur.

Carry out the same exercise for the layer between 90 kPa and 78 kPa.

3.9 Again considering the ascent in problem 3.7, take air from the surface up the dry adiabatic to its condensation level, then up the wet adiabatic until it reaches the temperature of the environment. By measuring the appropriate area on the diagram calculate the work required to raise 1 kg of air to this point.

Now continue the ascent along the wet adiabatic until the parcel again reaches the temperature of the environment. Calculate the energy released in this second ascent.

Show that for the complete ascent the net energy release is positive. Such a condition is known as *latent instability*. Once the surface has been heated sufficiently to initiate vertical motion, energy can be released and intense convective activity and maybe thunderstorms are likely to develop (cf. fig. 3.6).

3.10 Two parcels of damp air at the same pressure with masses M_1 and M_2, water vapour mixing ratios m_1 and m_2, and temperatures T_1 and T_2 are mixed. Show that the resulting temperature and mixing ratio are given by

$$\bar{T} = \frac{M_1 T_1 + M_2 T_2}{M_1 + M_2}$$

$$\bar{m} = \frac{M_1 m_1 + M_2 m_2}{M_1 + M_2}$$

(remember, $m \ll 1$).

Fig. 3.4. A visible image from the NOAA-8 satellite taken at 0919 GMT 25 April 1984 and received and processed by the University of Dundee Electronics Laboratory. Sea fog, which had formed over the eastern Atlantic as warm air moved over cooler water, is seen passing to the north of Scotland and entering the Moray Firth as winds turn to a more northerly point. The fog, although originally warmer than the sea, cooled by long-wave radiation until it became colder than the underlying sea surface. A vortex and lee waves (cf. §8.3) which formed downwind of the Faeroe Islands are detectable in the structure of the fog. Upper cloud associated with a high-level cold pool casts shadows on the fog top. Snow patches remain on the highest points of the Scottish highlands.

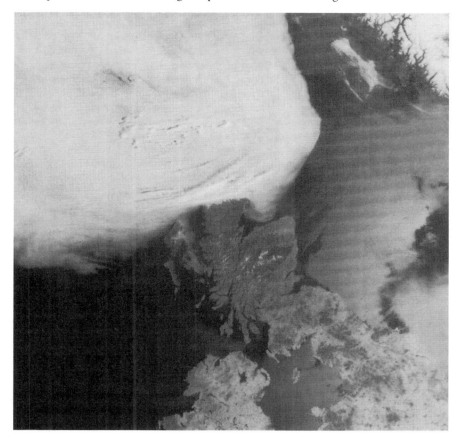

Using data available on the tephigram chart consider the mixing of two equal masses at 100 kPa pressure and 24°C and 12°C respectively. What relative humidity (assumed the same for each mass) must the masses possess for condensation to occur on mixing the masses together.

Fig. 3.5. Visible imagery obtained from a NOAA satellite at 1011 GMT on 14 February 1979 showing cloud lines associated with very cold continental air streaming westwards from Russia across the North Sea and British Isles. Parts of Norway and Denmark are visible in the east, while the British Isles can just be discerned beneath low-level convective cloud. With air temperatures over Denmark around −12°C and sea temperatures around 5°C, cumulus clouds formed rapidly over the sea in the gale-force Easterly winds. However, note the delayed formation of cloud to the lee of Norway where the air has been forced to descend. The depth of convection gradually increases as the air at low levels becomes warmer and moister, giving rise to larger cumulus cells within the bands and the formation of showers. The earlier formation of cloud cells over the Skagerrak (between Norway and Denmark) can be seen to produce a strip of more mature cumulus which may be traced right across the North Sea. Over the eastern half of Britain, where the air temperature was around −2°C, snow showers were widespread. Amounts of precipitation were small, however, because the cloud tops extended up to only 2000 m. To the west of the high ground of central Britain, the air dried out and much less convection is evident, although some snow showers occurred over Ireland. The main feature in the west is the large area of transverse lee waves (cf. §8.3) in the stratocumulus cloud which has formed beneath a marked inversion at 2 km height.

Around the southern portion of this image there are extensive layers of upper cloud associated with an old frontal system. Shadows can be seen along the northern edges of these layers, particularly across southern Denmark. (Courtesy of the Meteorological Office)

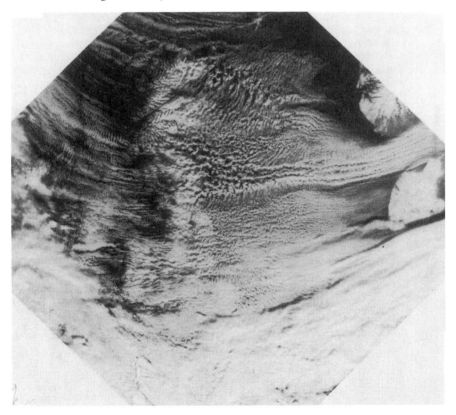

3.11 (1) Consider the vertical mixing of a column of air extending from pressure p_1 to pressure p_2. Show what the mixing ratio $\overline{m}$ of the mixed column is given by

$$\overline{m} = \frac{1}{p_2 - p_1} \int_{p_1}^{p_2} m \, dp$$

where $m(p)$ describes the initial distribution.

Show also with appropriate assumptions (state what they are) that an approximate expression for the potential temperature $\overline{\theta}$ of the mixed column is

$$\overline{\theta} = \frac{1}{p_2 - p_1} \int_{p_1}^{p_2} \theta \, dp$$

(2) Plot the following on a tephigram chart: 100 kPa, 28°C; 95 kPa, 24°C; 90 kPa, 21°C. The water vapour mixing ratios are 20 g kg^{-1} from 100 kPa to 95 kPa and 17 g kg^{-1} from 95 kPa to 90 kPa. If the 100 kPa to 90 kPa layer is mixed, estimate the resulting potential temperature and the condensation level. Your result demonstrates that mixing processes can lead to cloud or fog formation.

3.12 Plot on a tephigram chart the following large scale circulation cell between the equator and 30°N. Air rises adiabatically from 100 kPa and 25°C to 80 kPa when it becomes saturated and continues along the saturated adiabatic to 18.5 kPa and −75°C. It then moves northwards and sinks losing heat radiatively to 28 kPa and −72°C after which it descends adiabatically to the surface and returns along the surface while being heated to its starting point. Calculate the energy released in the cycle by unit mass of air.

If the overturning takes 6 months compare the rate of release of energy with the average input of solar radiation.

3.13 From the information contained in fig. 3.2 carry out a numerical integration of (3.22) and find the available potential energy in the northern hemisphere. Note that since fig. 3.2 contains zonally averaged information your answer refers to zonal available potential energy. Compare your result with that in fig. 3.3.

3.14 Derive (3.24) from (3.23) as follows:

(1)   If on an isobaric surface $\theta = \overline{\theta} + \theta'$, $\overline{\theta}$ being an average value over the surface, show that on a neighbouring isentropic surface where $p = \overline{p} + p'$,

$$p' = \theta' \, dp/d\theta$$

Fig. 3.6. (a) The cloud image (AVHRR visible channel received by University of Dundee Electronics Laboratory) was taken from the TIROS N satellite at 1428 GMT on 26 June 1980. The cloud seen over central and southern Britain and the near Continent is typical of that observed in a polar air mass on a summer's afternoon. Cold air passing over land warmed by the sun becomes unstable, generating updraughts and cloud. Over the neighbouring cooler sea, convection is largely absent.

Many of the clouds are bright, and therefore thick (cf. §6.4) with sharply defined edges; some can be seen to cast shadows on their north-eastern side. Some of the cloud systems extend over some tens of kilometres, while others are very small either because they are still growing or because their growth is inhibited by the compensating descent of air surrounding the vigorous updraughts of the larger systems.

(a)

Fig. 3.6. (*continued*) (b) The radar network picture (courtesy of the Meteorological Office) was obtained at a time corresponding to the satellite picture (radar techniques for precipitation measurement are described in §12.4). Many of the features seen on the satellite imagery can be seen to be correlated with the echoes from rainfall detectable by the radars. Note in particular the rain associated with the cloud systems just west of the Wash, over South Wales, around the Severn Estuary and over the south coast. Some of the rain was heavy, as depicted by white areas in the radar picture, and thunderstorms were reported. In the early stages of growth, the clouds consist of supercooled water droplets which freeze (glaciate) as the air rises. When the ice particles grow to a size such that their terminal velocities exceed the updraught velocity in the cloud, they fall to the ground, sweeping up supercooled droplets lower in the cloud. The smaller ice particles continue to be carried in the updraught, reach the upper troposphere and are blown downwind. These are responsible for characteristic 'anvil clouds' on a showery day and can be detected on the satellite picture as wispy cloud to the south-east of some of the bright echoes.

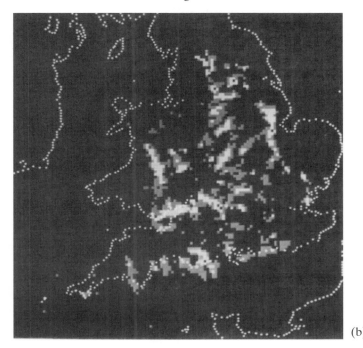

(b)

(2)   From (3.4) show that

$$\frac{d\theta}{dp} = -\kappa\theta p^{-1}\left(1 - \frac{\Gamma}{\Gamma_d}\right)$$

where $\Gamma = -dT/dz$ is the local lapse rate of temperature and $\Gamma_d = g/c_p$ is the adiabatic lapse rate.

(3)   Make the assumption that in the result of part (1) the average vertical stability may be employed, i.e. $p' = -\theta' d\overline{p}/d\theta$, and substitute from parts (1) and (2) in (3.23) to derive (3.24).

3.15   Compare the average input of solar radiation over the hemisphere with the following quantities in fig. 3.3:

(1)   rate of generation of zonal available potential energy,
(2)   rate of generation of kinetic energy.

3.16   From the information in fig. 3.2 plot a cross-section of the efficiency factor for generation of available potential energy (see (3.27)).

4

More complex radiation transfer

4.1 Solar radiation: its modification by scattering

In discussing radiation transfer in chapter 2 a very simplified atmospheric model was presented. The assumptions were made of an atmosphere in radiative equilibrium, transparent to solar radiation and possessing an absorption coefficient for long-wave radiation independent of frequency. In this chapter we shall consider how solar radiation is modified by the atmosphere and then set up the radiative transfer problem more generally showing how, for any given atmosphere, integrations over height and over frequency can be carried out.

Solar radiation is first modified by scattering processes within the atmosphere. A scattering coefficient σ can be defined in a similar way to the absorption coefficient k in equations (2.1) and (2.2) such that the transmission along an atmospheric path is given by

$$I = I_0 \exp(-\int \sigma \rho \, dz) \qquad (4.1)$$

Apart from scattering by clouds, which will be dealt with in chapter 6, the most important scattering mechanism is Rayleigh scattering by molecules, for which the scattering coefficient σ_R at wavelength λ is given by

$$\sigma_R = \frac{32\pi^3}{3N_0\lambda^4\rho_0}(n-1)^2 \qquad (4.2)$$

where N_0 is the number of molecules per unit volume, ρ_0 the density and n the refractive index, all at conditions of standard temperature and pressure. Attenuation by Rayleigh scattering varies very strongly with wavelength; note the λ^{-4} dependence of σ_R in equation (4.2). For a vertical column of atmosphere 40% or so is lost in the near ultraviolet while less than 1% is lost in the near infrared (problem 4.1). On average (taking into account different solar elevations and different wavelengths) about 13%

of the solar radiation incident on the atmosphere is Rayleigh scattered (problem 4.2). Approximately half of this reaches the earth's surface as *diffuse* radiation the other half being returned to space.

Solar radiation is also scattered by particulate aerosol arising from a wide variety of sources including volcanoes, chimneys, dust from the earth's surface and sea spray. Such aerosol is present in the atmosphere in very variable quantities (problem 4.3).

4.2   Absorption of solar radiation by ozone

Reference to fig. 2.1 and fig. A8.1 will indicate that, in addition to attenuation through the scattering processes discussed in §4.1 solar radiation is modified by absorption by atmospheric constituents – mainly by oxygen and ozone in the ultraviolet and by water vapour and carbon dioxide in the infrared. First we consider absorption of solar radiation by ozone in the stratosphere which is the major energy input to that region of the atmosphere.

Ozone is formed in the stratosphere (10–15 km) and mesosphere (50–80 km) by photochemical processes (§5.6), the peak of ozone concentration occurring at about 25 km altitude. Ozone absorbs ultraviolet solar radiation in the Hartley band (200–300 nm); at levels below 70 km virtually all the energy absorbed goes into kinetic energy of the molecules, i.e. into increasing the atmospheric temperature.

The absorption spectrum of ozone in the Hartley band is mainly a continuous spectrum, that is the absorption coefficient $k_{\tilde{\nu}}$ is a smoothly varying function of the wavenumber[†] $\tilde{\nu}$ and is also independent of pressure, so that, ignoring the complications of scattering which must be taken into account in an accurate calculation (cf. §12.8), the incident solar flux $F_{S\tilde{\nu}}(z)$ at height z is given by

$$F_{S\tilde{\nu}}(z) = F_{S\tilde{\nu}}(\infty)\exp\left(-\int_z^\infty k_{\tilde{\nu}}\rho_z \sec\theta \, dz\right) \qquad (4.3)$$

where ρ_z is the density of ozone at height z, and θ is the zenith angle of the sun.

Note that in order to take account of variations with wavenumber, the quantity $F_{S\tilde{\nu}}(z)$ is the solar flux per unit wavenumber interval at wavenumber $\tilde{\nu}$.

[†] Throughout this chapter, frequencies will be expressed in wavenumbers denoted by $\tilde{\nu}$. The wavenumber is the reciprocal of the wavelength or the frequency in Hz divided by the velocity of light.

To obtain the heating rate (cf. equation (2.4)) at level z an integration over frequency is required so that

$$c_p \rho \frac{dT}{dt} = \cos \theta \int_{\text{band}} \frac{dF_{S\tilde{\nu}}}{dz} d\tilde{\nu} \qquad (4.4)$$

Given values of $k_{\tilde{\nu}}$ across the spectrum and ρ_z at various heights, numerical integration of (4.3) and (4.4) can be carried out (see appendices 8 and 9 for information about solar flux and ozone absorption). For a typical ozone distribution at mid latitudes, the peak of heating is at ~50 km altitude; when integrated over a day, it amounts to a rate of change of temperature of ~8 K day^{-1} (problem 4.6).

4.3 Absorption by single lines

In the infrared part of the spectrum the absorption coefficient in molecular bands varies rapidly with frequency. Molecules possess discrete energy levels associated with different vibrational and rotational states; any given vibration-rotation band contains many thousands of individual lines (cf. fig. 4.1). The structure of the bands is often complex and is further complicated because each line in a band is broadened by collision broadening or Doppler broadening – broadening mechanisms which vary with temperature and pressure and which are, therefore, variable throughout the atmosphere.

In order to compute the average transmission of a path of atmosphere for a given spectral region, we need to sum over many individual lines.

The simplest model of a molecular band which may be employed is of a large number of lines which are independent (i.e. non-overlapping), broadening by collision processes only. For the carbon dioxide bands and water vapour bands of importance in the atmosphere this is a good approximation for levels in the stratosphere between ~20 km and 60 km altitude. Above these levels Doppler broadening becomes important (problem 4.7) and at lower levels lines overlap very substantially.

Simple theory leads to an absorption coefficient $k_{\tilde{\nu}}$ at wavenumber $\tilde{\nu}$ due to a collision broadened line centred at $\tilde{\nu}_0$ given by (see, for instance, Houghton & Smith, 1966; Eisberg, 1961)

$$k_{\tilde{\nu}} = \frac{s\gamma}{\pi \left\{ (\tilde{\nu} - \tilde{\nu}_0)^2 + \gamma^2 \right\}} \qquad (4.5)$$

where $s = \int_0^\infty k_{\tilde{\nu}} d\tilde{\nu}$ $\qquad (4.6)$

Fig. 4.1. Illustrating the fine structure of molecular absorption bands. Atmospheric transmission for a 10 km horizontal path at 12 km altitude as computed by McClatchey & Selby (1972) for (a) the 320–380 cm^{-1} (31–26 μm) region where the lines are due to water vapour, (b) the 680–740 cm^{-1} (15–13 μm) region where the lines are mainly due to carbon dioxide. Only a very small part of the spectrum is covered in this diagram. Compare with fig. 2.1 in which the variation of atmospheric transmission with frequency is on a very much coarser scale.

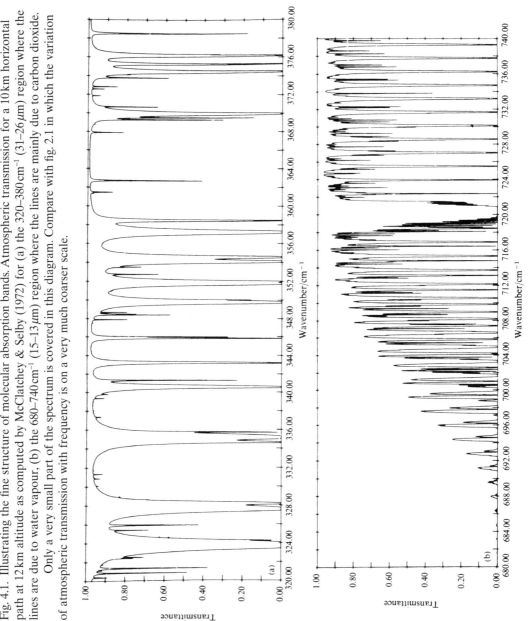

is the line strength and $\gamma = (2\pi tc)^{-1}$ the half-width (in wavenumbers). t is the mean time between collisions in the absorbing gas, a quantity which varies over several orders of magnitude in the atmosphere because it is inversely proportional to the pressure p. Ignoring the temperature dependence of γ which by comparison is small, we can write

$$\gamma = \gamma_0 p / p_0 \qquad (4.7)$$

where γ_0 is its value at standard pressure p_0. For many gases, at STP, $\gamma_0 \simeq 0.1\,\text{cm}^{-1}$, so that at 50 km altitude where the pressure is 0.1 kPa, $\gamma \simeq 10^{-4}\,\text{cm}^{-1}$.

Consider what appearance such a line would have when viewed with a high resolution laboratory spectrometer through a path of length l containing absorbing gas at density ρ. The transmission $\tau_{\tilde{\nu}}$ of such a path will be

$$\tau_{\tilde{\nu}} = \exp(-k_{\tilde{\nu}}\rho l) \qquad (4.8)$$

and the *equivalent width* or *integrated absorptance* W of the line given by fig. 4.2

$$\begin{aligned} W &= \int_0^{\infty} d\tilde{\nu}(1 - \tau_{\tilde{\nu}}) \\ &= \int_0^{\infty} d\tilde{\nu}\{1 - \exp(-k_{\tilde{\nu}}\rho l)\} \end{aligned} \qquad (4.9)$$

Two approximations for W are very useful: (1) the *weak* approximation where the exponential in (4.9) may be approximated by the first term of its expansion, in which case on substituting for $k_{\tilde{\nu}}$ from (4.5) (problem 4.8)

$$W = s\rho l \qquad (4.10)$$

and (2) the *strong* approximation for which there is no transmission of any consequence near the line centre (the line centre is then said to be black). The actual value of $k_{\tilde{\nu}}$ near the line centre is, therefore, not important providing it is large enough for γ^2 to be omitted from the denominator in (4.5). The integration in (4.9) may now be carried out (problem 4.9) and

$$W = 2(s\gamma\rho l)^{1/2} \qquad (4.11)$$

Plotting the equivalent width W against the length of absorbing path l results in the *curve of growth* (fig. 4.2).

Taking in a substantial proportion of an absorption band of width $\Delta\tilde{\nu}$ in which a number of non-overlapping lines are present, the average transmission $\bar{\tau}$ over the spectral interval will be

Fig. 4.2. (a) Curve of growth of a typical spectral line with $s = 10^4 \text{cm}^{-1} \,(\text{g cm}^{-2})^{-1}$ and $\gamma_0 = 0.06 \text{cm}^{-1}$ showing the linear and square root regions of growth.
(b) Actual shapes of the spectral line for different values of ρl corresponding to the values shown by the arrows in (a).

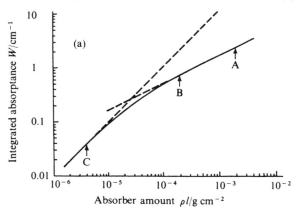

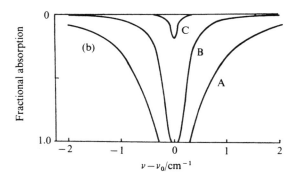

$$\bar{\tau} = 1 - \frac{\sum_i W_i}{\Delta \tilde{\nu}} \tag{4.12}$$

where W_i is the equivalent width of the ith line.

4.4   Transmission of an atmospheric path

We apply the results of the previous section to a vertical path in the atmosphere between levels of pressure p_1 and p_2 $(p_1 > p_2)$. When the weak approximation applies this is straightforward as the equivalent width is then only a function of the amount of absorber in the path, i.e.

$$W = s \int_{\text{path}} c\rho \, dz \tag{4.13}$$

where c is the fractional concentration (by mass) of absorber and ρ is the total density. Since $\rho \, dz = -dp/g$ and assuming a constant value of c along the path

$$W = sc(p_1 - p_2)/g \tag{4.14}$$

It is not so clear how to proceed when the absorption is stronger, as the pressure (and hence γ) varies along the path. In this case application is made of the *Curtis-Godson approximation* which states that a mean pressure $\bar{p}$ may be defined for a path

$$\bar{p} = \frac{\int pc\rho \, dz}{\int c\rho \, dz} \tag{4.15}$$

i.e. in deriving the mean pressure, the pressure along the path is weighted by the amount of absorber. This is an exact approximation when the strong approximation applies (problem 4.13); it is, of course, also exact under the weak approximation when the equivalent width is independent of pressure.

For uniform composition, $\bar{p} = \frac{1}{2}(p_1 + p_2)$ so that substituting from (4.7) and (4.15) in (4.11) we have in the strong approximation, for a single spectral line

$$W = 2\left\{\frac{s\gamma_0 c}{2gp_0}(p_1^2 - p_2^2)\right\}^{1/2} \tag{4.16}$$

in which case the average transmission over a spectral interval is, from (4.12) and (4.16)

$$\bar{\tau} = 1 - \frac{1}{\Delta\tilde{\nu}}\left\{\frac{2c}{gp_0}(p_1^2 - p_2^2)\right\}^{1/2}\sum_i (s_i\gamma_{0i})^{1/2} \tag{4.17}$$

where $\sum_i(s_i\gamma_{0i})^{1/2}$ is the sum of the square roots of the product of line strength and line width at STP within the spectral interval.

Provided the transmission is high the expression in (4.17) can be employed to work out the transmission for any given path or the proportion of solar radiation reaching different atmospheric levels (problem 4.15). A simple extension to the theory to cope with regions where lines overlap significantly will be presented in §4.9.

4.5   The integral equation of transfer

So far in this chapter we have dealt with the scattering of solar radiation and with its transmission in absorbing regions in the ultraviolet and the infrared parts of the spectrum. We now need to consider for the infrared part of the spectrum both emission and absorption. We follow on from the discussion of radiation transfer in chapter 2 where a very simplified atmospheric model was considered. There the assumptions were made of an atmosphere in radiative equilibrium and of an absorption coefficient independent of frequency. Here, we shall set up the problem more generally showing how, for any given atmosphere, integrations over height and over frequency can be carried out. We shall then consider a particular case, but still confining the discussion to absorption and emission processes only, i.e. to a non-scattering atmosphere.

To evaluate the effect of radiative transfer in the atmosphere we need to be able to find the intensity of radiation at any given level in an atmosphere with any structure and composition. For the plane parallel atmosphere considered in §2.2 an expression for the intensity is obtained by integrating Schwarzschild's equation (2.3). This may readily be done using the integrating factor $\exp(-\chi)$ (problem 4.16). The integral equation may also be derived directly in a way which illustrates the physical processes involved.

Note that because it is necessary to take into account variations of absorption coefficient with frequency (or wavenumber) we first consider the transfer of radiation of intensity $I_{\tilde{\nu}}$ at wavenumber $\tilde{\nu}$, the units of $I_{\tilde{\nu}}$ being Watts per unit area per unit solid angle per unit wavenumber interval.

Consider a thick slab (fig. 4.3) between levels z_0 and z_1, with radiant intensity $I_{\tilde{\nu}0}$ incident vertically upwards at z_0. To calculate $I_{\tilde{\nu}1}$ the intensity

Fig. 4.3

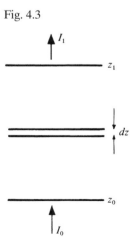

leaving the top of the slab, consider the elemental slab dz at level z ($z_0 < z < z_1$) and temperature T emitting radiation in the upward direction of intensity $k_{\tilde{\nu}} \rho\, dz\, B_{\tilde{\nu}}(z)$ (cf. §2.2) where $k_{\tilde{\nu}}$ is the absorption coefficient at wavenumber $\tilde{\nu}$ and

$$B_{\tilde{\nu}} = 2\tilde{\nu}^2 \frac{hc\tilde{\nu}}{\exp(hc\tilde{\nu}/kT) - 1} \tag{4.18}$$

is the Planck function at wavenumber $\tilde{\nu}$ (cf. appendix 7), where the definition of the quantities is as in §2.2. This emitted radiation will be attenuated before reaching z_1, the transmission (§2.2) being

$$\tau_{\tilde{\nu}}(z, z_1) = \exp\left(-\int_z^{z_1} k_{\tilde{\nu}} \rho\, dz'\right) \tag{4.19}$$

The contribution to $I_{\tilde{\nu}1}$ from the slab dz will be

$$\begin{aligned} dI_{\tilde{\nu}1} &= k_{\tilde{\nu}} \rho\, dz\, B_{\tilde{\nu}}(z) \exp\left(-\int_z^{z_1} k_{\tilde{\nu}} \rho\, dz'\right) \\ &= B_{\tilde{\nu}}(z)\, d\tau_{\tilde{\nu}}(z, z_1) \end{aligned} \tag{4.20}$$

as may easily be verified by differentiating (4.19). We have, therefore,

$$I_{\tilde{\nu}1} = I_{\tilde{\nu}0}\tau_{\tilde{\nu}}(z_0, z_1) + \int_{\tau_{\tilde{\nu}}(z_0, z_1)}^{1} B_{\tilde{\nu}}(z)\, d\tau_{\tilde{\nu}}(z, z_1) \tag{4.21}$$

the first term being the contribution from the incident intensity at z_0.

Equation (4.21) is the integral equation of transfer and is important not only in radiation transfer within the atmosphere but also in the interpretation in terms of atmospheric structure of measurements of radiation leaving the top of the atmosphere (§12.6).

4.6   Integration over frequency

To obtain the total intensity of radiation at any given atmospheric level, it is necessary to integrate over all frequencies where significant radiation is present. Integrating (4.21) over frequency leads to

$$I_1 = \int_0^\infty I_{\tilde{\nu}0}\tau_{\tilde{\nu}}(z_0, z_1)d\tilde{\nu} + \int_0^\infty \int_{\tau_{\tilde{\nu}}(z_0, z_1)}^{1} B_{\tilde{\nu}}(z)\, d\tau_{\tilde{\nu}}(z, z_1)d\tilde{\nu} \tag{4.22}$$

Of the quantities on the right hand side of (4.22), $B_{\tilde{\nu}}$ is a relatively slowly varying function of frequency whereas $\tau_{\tilde{\nu}}$ varies very rapidly with frequency (fig. 4.1). Given adequate information about the intensities, positions, widths and shapes of the spectral lines involved in any given spectral region, it is possible in principle to perform the integration over frequency in (4.22). However, to cope in practice with the integration it is necessary to find simplified methods of removing much of the complication associated with molecular band structure. One such simplified method employing the non-overlapping line approximation was discussed in §4.3; further methods will be presented in §4.9 and problem 4.19.

4.7   Heating rate due to radiative processes

The main reason why radiative transfer calculations are required is to deduce the contribution to the atmosphere's energy budget from radiative processes. The local heating rate is expressed in (2.4) in terms of the divergence of the upward and downward radiation fluxes.

For nearly all atmospheric situations it is adequate to employ the simple approximation of §2.2 for integration over angle, so that if in (4.22) τ is replaced by τ^* (i.e. path lengths are multiplied by $\frac{5}{3}$) and B is replaced by πB, an equation for the upward flux $F^\uparrow$ at level z_1 can be written

$$F^\uparrow = \int_0^\infty F_{\tilde{\nu}0} \tau_{\tilde{\nu}}^*(z_0, z_1)\, d\tilde{\nu} + \int_0^\infty \int_{\tau_{\tilde{\nu}}^*(z_0,z_1)}^1 \pi B_{\tilde{\nu}}(z)\, d\tau_{\tilde{\nu}}^*(z, z_1)\, d\tilde{\nu} \tag{4.23}$$

An entirely similar expression applies for the downward flux $F^\downarrow$.

4.8   Cooling by carbon dioxide emission from upper stratosphere and lower mesosphere

We now apply the non-overlapping strong collision-broadened line approximation to the ν_2 band of carbon dioxide at $15\,\mu m$ wavelength in the 20–60 km region. In this part of the atmosphere the temperature distribution is mainly determined by a balance between radiative cooling in the infrared from this carbon dioxide band and heating by absorption of solar radiation by ozone (§4.2).

To compute the radiation loss and hence the cooling rate we shall employ the *cooling to space* approximation, in which exchange of radiation between different atmospheric layers is neglected in comparison with the loss of radiation direct to space – a good approximation for this particular atmospheric region. Equating the loss of energy to space, the term $dF^\uparrow/dz$ from (2.4), to the cooling rate we have

$$\frac{dT}{dt}\rho c_p = -\int_{\Delta\tilde{\nu}} \pi B_{\tilde{\nu}}(T)\frac{d\tau_{\tilde{\nu}}^*(z, \infty)}{dz}\, d\tilde{\nu} \tag{4.24}$$

where the integration over frequency is as in (4.23) and covers the region $\Delta\tilde{\nu}$ of the $15\,\mu m$ carbon dioxide band. Over this spectral interval $B_{\tilde{\nu}}(T)$ can be considered constant with frequency. From (4.17)

$$\int_{\Delta\tilde{\nu}} \tau_{\tilde{\nu}}^*(z, \infty)\, d\tilde{\nu} = \Delta\tilde{\nu} - p\left(\frac{10c}{3gp_0}\right)^{1/2} \sum_i (s_i \gamma_{0i})^{1/2} \qquad (4.25)$$

The factor $\frac{5}{3}$ is included because we require the slab transmission appropriate to flux calculations. The strong approximation has also been applied to all the lines in the band; the result of problem 4.20 shows that this is a valid simplification. Substituting in (4.24) and remembering that $\rho\, dz = -dp/g$ we have

$$\frac{dT}{dt} = -\frac{g\pi B_{\tilde{\nu}}(T)}{c_p}\left(\frac{10c}{3gp_0}\right)^{1/2} \sum_i (s_i \gamma_{0i})^{1/2} \qquad (4.26)$$

Substituting values $c_p = 1005\,\mathrm{J\,kg^{-1}}$, $c = 5.5 \times 10^{-4}$ (360 ppm), $\Sigma_i(s_i\gamma_{0i})^{1/2} = 1600\,\mathrm{cm^{-1}\,(g\,cm^{-2})^{-1/2}}$ (appendix 10), we have

$$\frac{dT}{dt} = 2.16\pi B_{\tilde{\nu}}\,\mathrm{K\,s^{-1}} \qquad (4.27)$$

where $\pi B_{\tilde{\nu}}$ is in $\mathrm{W\,cm^{-2}\,(cm^{-1})^{-1}}$. Note the important result that the factor multiplying $B_{\tilde{\nu}}$ in the expression for the cooling rate is independent of height.

A first approximation to the temperature near 50 km altitude may be obtained by assuming that radiative equilibrium applies and that the heating due to the absorption of solar radiation by ozone (§4.2) is balanced by the cooling due to CO_2 emission. Inserting the ozone heating rate from problem 4.6 in equation (4.27), the equilibrium value of $\pi B_{\tilde{\nu}}(T)$ at 50 km is found to be $\sim 3.8 \times 10^{-5}\,\mathrm{W\,cm^{-2}\,(cm^{-1})^{-1}}$ equivalent to a temperature T of $\sim 280\,\mathrm{K}$, approximately as found at that level. A diurnal temperature variation of $\sim 4\,\mathrm{K}$ is present at 50 km because heating only occurs during the day but cooling goes on all the time.

4.9 Band models

The simplest model of a band is that of non-overlapping lines described in §4.3. Such a model, however, is only applicable when the lines are well separated. For the atmosphere it can only be applied for the path lengths and pressures above altitudes of about 30 km. A variety of band models have been invented which allow for overlap, in particular the regular model first described by Elsasser, in which an absorption band is simulated by an array of equally spaced lines of the same shape and strength, and the random model due to Goody, in which the lines are considered to be spaced randomly and to have a distribution of line strengths following some statistical law. This latter model leads to a particularly simple and useful expression for the average transmission $\bar{\tau}$ of a spectral interval of width $\Delta\tilde{\nu}$ containing a large number of lines

$$\bar{\tau} = \exp(-\sum W_i / \Delta\tilde{\nu}) \qquad (4.28)$$

where $\sum W_i$ is the sum of the equivalent widths of all the lines in the interval considered as independent lines. In problem 4.19 the theory of the random band model is developed further. From this theory and from the spectral data provided in appendix 10, the transmission of any spectral interval in the bands of water vapour, carbon dioxide and ozone in the region of the thermal infrared can be calculated.

With the use of band models, the frequency integration of (4.22) may be carried out for rather few spectral intervals, each major absorption band being divided into a few spectral regions. Such simplified radiation calculations are required for incorporation into numerical models (§11.7).

4.10 Continuum absorption

In the wavelength region $8-13\,\mu$m in the infrared, apart from the ozone band at $9.6\,\mu$m, the atmosphere is reasonably transparent (fig. 2.1). Some absorption by water vapour lines is still present, however, and of more importance, by water vapour dimers, i.e. associations of water vapour molecules in pairs. This latter absorption is a continuum with absorption coefficient proportional to the partial pressure of water vapour; associations are more likely to occur when the partial pressure is high. For this reason the continuum absorption is particularly important in humid atmospheres; for a wet tropical atmosphere, transmission of a vertical path through the atmosphere may only be ~50% in this region (problem 4.17).

4.11 Global radiation budget

In this chapter, equations have been written down which, given detailed information regarding the spectra of the molecules, enable calculations of the radiation fluxes and the radiation heating rate to be made at any level in a clear atmosphere. No account has been taken of clouds and their radiative properties. Clouds are, in fact, probably the dominant influence in the radiative budget of the lower atmosphere but adequately taking them into account raises many problems, which will be considered briefly in §6.4.

Fig. 4.4. The earth's radiation and energy balance (after Kiehl & Trenberth, 1997). The net incoming solar radiation of 342 W m^{-2} is partially reflected by clouds and the atmosphere, or at the surface; 49% is absorbed by the surface. That heat is returned to the atmosphere, some as sensible heat but most from evaporation and transpiration (these taken together are known as evapotranspiration) that is later released as latent heat of condensation. The rest is radiated as thermal infrared radiation from the surface much of which is absorbed by the atmosphere or by clouds which also emit radiation both upwards and downwards. The infrared radiation that is lost to space comes from cloud tops and parts of the atmosphere much colder than the surface.

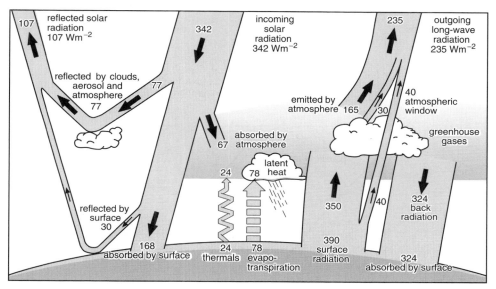

Of fundamental importance to our understanding of the atmosphere's energy budget is adequate knowledge of the components of the radiation budget at the top of the atmosphere. Fig. 4.4 shows what happens on average to the incident solar radiation. It also shows that the outgoing long-wave radiation originates from near the surface or from cloud tops in the window region and from high in the troposphere in the strong carbon dioxide and water vapour bands (cf. fig. 12.7). Fig. 4.5 shows the results of satellite measurements of the two main components of the radiation budget averaged over longitude and over a complete year. The excess of solar energy absorbed over that emitted in equatorial regions must be transported through atmospheric motions or ocean currents to the polar regions where it is required to supply the deficit between energy emitted and received. How atmospheric motions achieve the transfer will be discussed in chapter 10.

Fig. 4.5. Average components of the earth's radiation budget as deduced from satellite observations, 1962–66, by Vonder Haar & Suomi (1971).

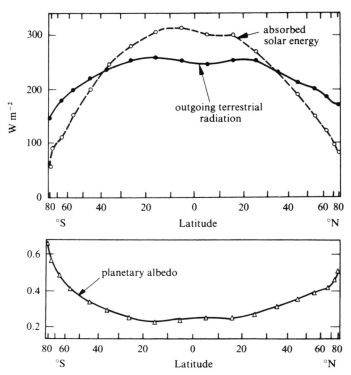

Problems

4.1 From the expression (4.2) for the scattering coefficient for Rayleigh scattering at wavelength λ, and using the data in appendix 1, calculate the extinction by Rayleigh scattering for a vertical column of atmosphere at wavelengths of 0.3, 0.6 and 1 μm.

4.2 Continue the calculations of problem 4.1 for a few other wavelengths and for a number of solar elevations. Hence estimate, for a clear atmosphere, with Rayleigh scattering only, the proportion of solar radiation scattered out by the atmosphere from the whole hemisphere illuminated by the sun.

4.3 The Angström turbidity factor of an atmosphere containing aerosol is defined as the optical thickness τ_A of a vertical column due to aerosol scattering. Assuming that the scattering cross-section of the particles is twice their geometric cross-section, estimate what density of aerosol of diameter 1 μm would be needed in the lowest kilometre of the atmosphere to produce a turbidity factor of unity.

4.4 It is commonly assumed that the Angström turbidity factor varies with wavelength λ as $\lambda^{-1.3}$. For an atmosphere with an optical depth due to aerosol of 0.3 at $0.6\,\mu$m, what is the value of the turbidity coefficient at $0.3\,\mu$m and $1\,\mu$m respectively. From the results of problem 4.1, calculate the total extinction due to Rayleigh and aerosol scattering for such an atmosphere at the three wavelengths.

4.5 Explain why scattered sunlight observed at 90° to the direction of the sun shows very strong polarization. Why is the polarization only about 90%, not 100%?

4.6 Assuming the sun overhead and a uniform temperature atmosphere, for a distribution of the number density n_3 of ozone molecules as a function of pressure p given by $n_3 = n_0 p^{3/2}$ (quite a good approximation to the upper part of the ozone layer), show that the heating rate h by absorption of solar radiation is of the form

$$\frac{h}{h_m} = \left(\frac{p}{p_m}\right)^{1/2} \exp\left\{-\frac{1}{3}\left[\left(\frac{p}{p_m}\right)^{3/2} - 1\right]\right\} \tag{4.29}$$

where p_m is the level of maximum heating h_m. In your calculation assume a single absorption coefficient independent of wavelength within the ozone band.

Given that the level of maximum heating is at $0.1\,$kPa pressure, the solar flux over the ozone absorbing region is $0.7 \times 10^{-3}\,$W cm^{-2}, calculate h_m.

4.7 Because of the ellipticity of the earth's orbit the solar radiation incident on the earth is about 7% greater in December than in June. Assuming constant ozone, from the variation with temperature of the Planck function at $15\,\mu$m, estimate the temperature difference that arises for this reason between the temperature of the stratopause (the level of maximum temperature near $50\,$km) in the summer over the north pole and over the south pole.

4.8 The shape of a spectral line due to Doppler broadening is given by

$$k_{\tilde{\nu}} = \frac{s}{\gamma_D \pi^{1/2}} \exp\left\{-\left(\frac{\tilde{\nu} - \tilde{\nu}_0}{\gamma_D}\right)^2\right\} \tag{4.30}$$

where $\gamma_D = \frac{\tilde{\nu}_0}{c}\left(\frac{2RT}{M_r}\right)^{1/2}$ \hfill (4.31)

(M_r is molecular weight).

Calculate the pressure in the atmosphere at which the half-width due to collision broadening is equal to that due to Doppler broadening for (1) a carbon dioxide line at $667\,\mathrm{cm}^{-1}$, (2) a water vapour line at $1600\,\mathrm{cm}^{-1}$, (3) a water vapour line at $100\,\mathrm{cm}^{-1}$. Assume all lines have collision broadened half-widths of $0.1\,\mathrm{cm}^{-1}$ at standard pressure.

For the two shapes compare the ratio of the absorption coefficient at the line centre to its value several half-widths away. Notice that the absorption coefficient in the wings of collision broadened lines is relatively much greater than that for the wings of Doppler broadened lines. For this reason, even for conditions where the Doppler half-width is greater than the collision broadened half-width the transfer of radiation in the collision broadened wings of the lines remains important.

4.9   Integrate (4.9) under the weak approximation to obtain (4.10).

4.10 Integrate (4.9) under the strong approximation to obtain (4.11).

4.11 Consider a path of fixed length l in which the pressure of an absorbing gas can be varied. Show that under conditions of collision broadening the absorption at the line centre is independent of pressure.

4.12 Show that for a single collision broadened line the integral of (4.9) may be written as

$$W = \int_{-\infty}^{\infty} dx \left\{ 1 - \exp\left(\frac{-2u\gamma^2}{x^2 + \gamma^2} \right) \right\}$$

where $x = \tilde{\nu} - \tilde{\nu}_0, \quad u = s\rho l / 2\pi\gamma$

The value of this integral may be expressed in terms of the modified Bessel functions $I_n(u) = -iJ_n(iu)$ giving

$$W = 2\pi\gamma l u \exp(-u)\{J_0(u) + I_1(u)\}$$
$$= 2\pi\gamma L(u) \tag{4.32}$$

The function $L(u)$ is known as the *Ladenberg and Reiche function*.

4.13 Substitute the strong approximation to $k_{\tilde{\nu}}$ from (4.5) (i.e. omit the γ^2 in the denominator) into (4.19) for the transmission along an atmospheric path and show that the Curtis–Godson approximation (4.15) is exact in the case of the strong approximation.

4.14 Two useful approximations to the equivalent width W of a collision broadened line are

$$W = s\rho l \left(1 + \frac{s\rho l}{4\gamma}\right)^{-1/2}$$

and

$$W = s\rho l \left\{1 + \left(\frac{s\rho l}{4\gamma}\right)^{5/4}\right\}^{-2/5}$$

Show that these expressions have the correct 'weak' and 'strong' limits ((4.10) and (4.11)). The first deviates by less than 8% from the correct equivalent width for all values of the parameters. The second due to Goldman (1968) has a maximum error of about 1%.

4.15 From equations (4.17) and (4.28) and the data of appendices 2, 3 and 10 estimate the average transmission of a vertical path from the top of the atmosphere to the 20 kPa level for the spectral interval 525–550 cm^{-1}. Also for latitude 60°N estimate the transmission of a vertical path of atmosphere down to the surface for the spectral interval 775–800 cm^{-1}.

4.16 Integrate (2.3) to derive (4.21) using the integrating factor $\exp(-\chi)$.

4.17 Absorption in the atmospheric window between 8 μm and 13 μm is mostly due to the water vapour dimer; the absorption coefficient is of the form $k_2 e$ where e is the water vapour pressure (in kPa), $k_2 \simeq 10^{-1}$ (g cm^{-2})$^{-1}$ kPa^{-1}. If the water vapour pressure near the surface is 1 kPa, calculate (1) the transmission of a horizontal path 1 km long near the surface, (2) the transmission of a vertical path of atmosphere assuming that the distribution of water vapour pressure is proportional to (pressure in atmospheres)^4. Also estimate for a layer near the surface the cooling rate in K day^{-1} by radiation from water vapour in this spectral region.

4.18 Integrate (4.21) by parts to give

$$I_{\tilde{\nu}1} = \{I_{\tilde{\nu}0} - B_{\tilde{\nu}}(z_0)\}\tau_{\tilde{\nu}}(z_0, z_1) + B_{\tilde{\nu}}(z_1) + \int_{z_1}^{z_0} \tau_{\tilde{\nu}}(z, z_1)\frac{dB_{\tilde{\nu}}(z)}{dz}dz$$

Write this equation in terms of flux and using (2.4) show that when $I_{\tilde{\nu}0} = B_{\tilde{\nu}}(z_0)$ an expression for the atmospheric heating rate at level z_1 is

$$\rho c_p \frac{dT}{dt} = \pi \int_0^\infty \int_0^\infty \frac{d\tau_{\tilde{\nu}}(z, z_1)}{dz_1} \cdot \frac{dB_{\tilde{\nu}}(z)}{dz}dz\, d\tilde{\nu}$$

4.19 Suppose in a random array of spectral lines the number $N(s)ds$ of lines having a strength between s and $s + ds$ is

$$N(s)ds = \frac{N_0}{s}\exp\left(-\frac{s}{\sigma}\right) \tag{4.33}$$

This distribution is known as the Malkmus model (Malkmus, 1967) and fits well to line distributions found in practice. Show that the sum of the equivalent widths ΣW_i of these lines (assuming no overlap) after passing through a path of length l with absorber density ρ is

$$\sum W_i = \int_0^\infty \frac{N_0}{s}\exp\left(-\frac{s}{\sigma}\right)ds\int_0^\infty \{1 - \exp(-k_{\tilde{\nu}}\rho l)\}d\tilde{\nu} \tag{4.34}$$

If a line shape is described by

$$k_{\tilde{\nu}} = sf(\tilde{\nu}) \tag{4.35}$$

carry out the integration over s to give

$$\sum W_i = N_0\int_0^\infty \ln\{1 + \sigma\rho lf(\tilde{\nu})\}d\tilde{\nu} \tag{4.36}$$

If $k_{\tilde{\nu}}$ is given by the collision broadened expression (4.5) show that

$$\sum W_i = 2\pi\gamma N_0\left\{\left(1 + \frac{\sigma\rho l}{\pi\gamma}\right)^{1/2} - 1\right\} \tag{4.37}$$

Compare the values of ΣW_i given by (4.37) in the limits of $\sigma\rho l/\pi\gamma \ll 1$ and $\gg 1$ with what would be expected from the weak and strong approximations (4.10) and (4.11) respectively in

$$\sum_i W_i(\text{weak}) = \sum_i s_i\rho l \tag{4.38}$$

$$\sum_i W_i(\text{strong}) = 2\sum_i (s_i\gamma_i\rho l)^{1/2} \tag{4.39}$$

where the summation in each case is over all the lines denoted by strength s_i and half-width γ_i.

The best way of fitting spectral line data to the random model is not by attempting to match the model distribution of line strengths with the actual distribution but rather by matching the weak and strong limits of (4.37) with (4.38) and (4.39) respectively. By doing this, find expressions for $\pi N_0\gamma$ and $\sigma/\pi\gamma$ in (4.37) in terms of Σs_i and $\Sigma(s_i\gamma_i)^{1/2}$. Hence show that (4.37) may be written

$$\sum W_i = \frac{2\left\{\sum (s_i\gamma_i)^{1/2}\right\}^2}{\sum s_i}\left[\left\{1 + \frac{\rho l(\sum s_i)^2}{\left[\sum (s_i\gamma_i)^{1/2}\right]^2}\right\}^{1/2} - 1\right] \tag{4.40}$$

This is a particularly useful form of ΣW_i which, for any given spectral interval $\Delta\tilde{\nu}$, can be substituted in (4.28) to calculate the transmission under any given circumstance when collision broadening applies. Spectral line data for many bands of importance suitable for substituting in (4.40) are listed in appendix 10.

4.20 For calculations of the transmission of a vertical path of atmosphere between a level where the pressure is p and the top of the atmosphere, show, by applying the data of appendix 10 for the $15\,\mu$m carbon dioxide band to (4.40) that, providing collision broadening applies, the strong approximation is a very good assumption.

5

The middle and upper atmospheres

5.1   Temperature structure

Fig. 5.1 shows a typical cross-section of the temperature structure of the earth's atmosphere from the surface up to ~100 km. The temperature falls with increasing height in the lowest 10 or 15 km roughly at the adiabatic lapse rate, as discussed in chapter 2. Above the tropopause the main features of the temperature structure are determined by radiative processes. The high temperature at ~50 km (known as the *stratopause*) is due to absorption of solar radiation by ozone (§4.2) – balanced mainly by emission of infrared radiation by carbon dioxide. Above the temperature peak at 50 km is the *mesosphere* (middle sphere) where the temperature falls with height, although not so steeply as in the troposphere. The upper boundary of the mesosphere is the temperature minimum around 80 km altitude known as the *mesopause* where there is rather little absorption of solar radiation. Above the mesopause, solar ultraviolet radiation is strongly absorbed, particularly by molecular and atomic oxygen, and the temperature rises rapidly to between 500 K and 2000 K in the region known as the *thermosphere*.

The variation of temperature with latitude should be noted from fig. 5.1. At the earth's surface and at the level of the stratopause the equator is warmer than the polar regions, as simple considerations would predict. The tropopause and the mesopause are, however, substantially colder over the equator than over the poles. This is particularly surprising for the winter polar regions where there is no solar radiation. Radiative considerations cannot provide an explanation of these reversed temperature gradients. Their existence demonstrates that atmospheric motions not only act as heat engines and transfer energy from source regions to sink regions but can also act as refrigerators transferring energy in the other direction.

Fig. 5.1. Mean latitudinal cross-sections, surface to 120 km, of (a) temperature (K) and (b) zonal wind (ms^{-1}), for March, June, September, and December. The relation between temperature and wind fields is described in §7.6. Data for the diagrams are tabulated in Appendix 5 where the sources of data are listed.

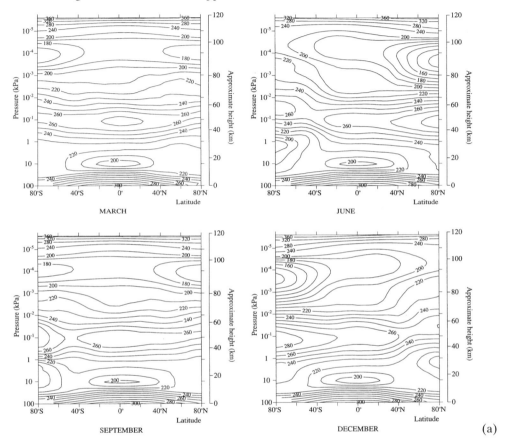

(a)

In the following sections of this chapter we shall consider various physical processes which occur in the atmosphere above 30 km. Only 1% of the atmospheric mass lies above 30 km and only 1 part in 10^5 above ~80 km. It must not be expected, therefore, that events in these upper regions have a very immediate or large effect on the structure or motion of the troposphere. However, because of the close links which exist between motions in various parts of the atmosphere, study of upper atmospheric motion can have an important bearing on our understanding of motion lower down. Further, what happens at the top of the atmosphere is important in the study of the long-term evolution of the atmosphere. Consideration, therefore, of upper atmosphere phenomena, in addition to

Fig. 5.1. (*continued*)

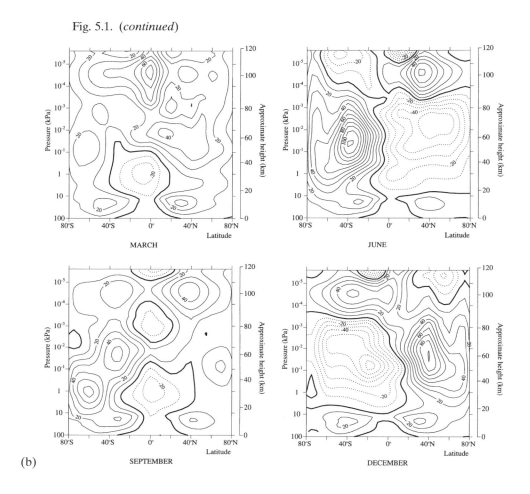

(b)

being of interest in its own right, plays an important part in the understanding of the whole.

5.2   Diffusive separation

The amount of mixing in the lower atmosphere ensures that no
significant diffusive separation occurs between the heavy atmospheric
constituents (e.g. argon) and the light ones (e.g. hydrogen). For practical
purposes, therefore, the main constituents of the lower atmosphere are uniformly mixed apart from those which are involved in phase changes (e.g.
water vapour) or chemical changes (e.g. ozone).

Although turbulent mixing dominates over molecular diffusion
in the lower atmosphere, because the molecular diffusion coefficient is
proportional to the mean free path and hence inversely proportional to

density, in the very high atmosphere molecular diffusion plays the dominant role both for the transfer of the molecules themselves and for the transfer of heat. A typical value of molecular diffusion coefficient for an atmospheric gas is $\sim 10^{-4}\,m^2\,s^{-1}$ at STP and therefore $\sim 100\,m^2\,s^{-1}$ at a pressure of 10^{-6} atm (corresponding to ~ 96 km altitude). This is of the same order as a typical eddy diffusion coefficient (cf. §9.2) appropriate to the lower thermosphere. The lower boundary of the region above which molecular diffusion dominates is ~ 120 km altitude and is known as the *turbopause* (sometimes known as the *homopause*: the region above, where molecular diffusion is dominant, is sometimes known as the *homosphere*). The turbopause can often be observed fairly clearly from the luminous trails of rockets, as a boundary below which the trail is violently disturbed by turbulence and above which it is relatively smooth.

We first consider the composition of an atmosphere in a state of diffusive equilibrium. In such equilibrium the hydrostatic equation (1.4) is applied to each constituent separately. This follows from Dalton's law of partial pressures, together with the condition that there be no net transport of any constituent across a horizontal surface. For the case of an isothermal atmosphere

$$\ln\{p(i)/p_0(i)\} = -z/H(i) \tag{5.1}$$

where the index i refers to the ith constituent. If $T = 1000$ K, the order of temperature found in the thermosphere, the values of the scale height H for argon ($M_r = 40$), nitrogen ($M_r = 28$), atomic oxygen ($M_r = 16$), and atomic hydrogen ($M_r = 1$) are respectively 21 km, 30 km, 54 km and 850 km. From (5.1), the ratio of the number densities n of two different constituents at an altitude z may be compared to their number densities n_0 at altitude z_0 at which diffusive separation begins. For example, for argon and nitrogen,

$$\frac{n(A)n_0(N_2)}{n_0(A)n(N_2)} = \exp\left[-(z - z_0)\left\{\frac{1}{H(A)} - \frac{1}{H(N_2)}\right\}\right] \tag{5.2}$$

The results of such calculations are shown in fig. 5.2. Whereas nitrogen is the most abundant constituent in the lower atmosphere, this role is taken over by atomic oxygen at about 170 km and by helium at about 500 km, these altitudes being quoted for mean atmospheric conditions; they vary very considerably with atmospheric temperature (problem 5.1).

5.3 The escape of hydrogen

As density decreases in the atmosphere with increasing altitude, eventually a level is reached where collisions are so rare that a molecule

Fig. 5.2. Total number density and number densities of various constituents for a mean atmosphere. (From COSPAR, 1972)

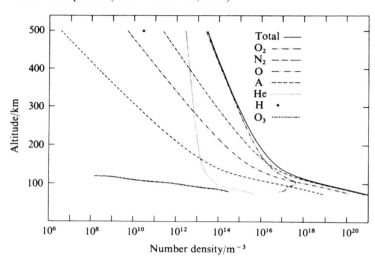

moving upwards will turn about under the influence of gravity and return again to the denser layers without making a collision. Under such conditions the faster molecules escape from the atmosphere altogether. The region is called the *exosphere*. The *critical level* z_c for such escape may be defined such that a proportion e^{-1} of a group of very fast particles moving vertically upwards at z_c will experience no collisions as they pass completely out of the atmosphere. For molecules of diameter d, the probability of no collision being made in a vertical path dz is $\exp[-\pi d^2 n(z)dz]$, where $n(z)$ is the particle density at height z. Substituting for $n(z)$ from the hydrostatic equation (1.4) the condition for the critical level z_c is therefore

$$\int_{z_c}^{\infty} \pi d^2 n(z_c) \exp\left[\frac{-(z - z_c)}{H}\right] dz = 1 \qquad (5.3)$$

i.e. $n(z_c) = \left(\pi d^2 H\right)^{-1}$ \qquad\qquad\qquad\qquad\qquad (5.4)

The departure from equilibrium is sufficiently small for the scale height H to be considered approximately constant through the region. Since $[n(z)\pi d^2]^{-1}$ is the mean free path in the horizontal direction at level z, the critical level may also be defined as that at which the mean free path in the horizontal direction is equal to the scale height H. The critical level occurs at an altitude where the most abundant constituent is atomic oxygen (fig. 5.2). For a temperature of 1000 K (approximate value for mean solar activity (fig. 5.3)) its scale height H is 53 km. With $d \simeq 2 \times 10^{-10}$ m,

Fig. 5.3. (a) Mean temperature of the upper atmosphere during periods of (a) very low, (b) mean, (c) very high solar activity; (b) the variation with local time of exospheric temperature over the equator at the equinox. (From COSPAR 1972; for more information see COSPAR 1988)

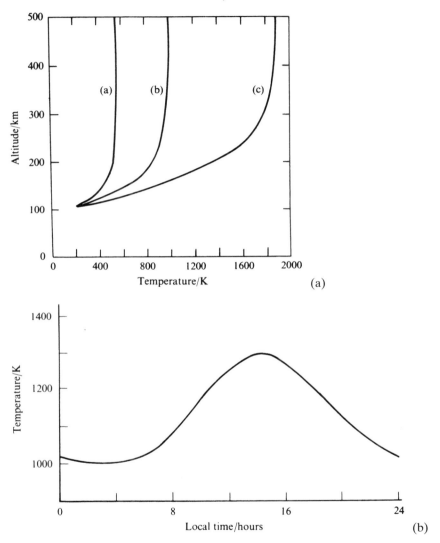

$n(z_c) \simeq 10^{14}\,\mathrm{m}^{-3}$ corresponding to a height z_c of between 400 and 500 km (fig. 5.2). Considerable variations, of course, occur with solar activity, time of day and season.

We now consider the rate of escape of hydrogen atoms. These originate mainly from molecules of water vapour and methane which are dissociated in the stratosphere and mesosphere by the action of ultraviolet radiation from the sun. For escape upwards from the earth's gravitational field a particle of mass m and velocity V must possess a kinetic energy $\frac{1}{2}mV^2$ greater than its gravitational potential energy mGM/a (G is the gravitational constant and M and a respectively the mass and radius of the earth), i.e.

$$\frac{1}{2}mV^2 > \frac{mGM}{a}$$

or $V^2 > 2ga, \quad \text{since } g = \frac{GM}{a^2}$ (5.5)

or $V > V_c$

where $V_c = 11\,\mathrm{km\,s^{-1}}$

This is over twice the mean molecular velocity of atomic hydrogen at a temperature of 1000 K, and very much larger than the mean velocities of other molecules.

Since the rate of escape is small the velocity distribution at the escape level will be approximately Boltzmann. The number dn_H of hydrogen atoms per unit volume having velocities between V and $V + dV$ is (see, for instance, Jeans, 1940)

$$dn_H = \frac{n_H(z_c)V^2 dV \exp(-\alpha V^2)}{\int_0^\infty V^2 dV \exp(-\alpha V^2)}$$ (5.6)

where $\alpha = m/2kT$ and m is the mass of a hydrogen atom. Of these the number moving in an upward direction across unit area per second is $\frac{1}{4}V dn_H$ (see, for instance, Jeans, 1940), so that the number $\dot{N}$ of atoms escaping per unit time from unit area is

$$\dot{N} = \frac{1}{4}n_H(z_c)\left(\frac{\alpha^3}{\pi}\right)^{1/2}\int_{V_c}^\infty V^3 dV \exp(-\alpha V^2)\,dV$$ (5.7)

where a further factor $\frac{1}{4}$ has been included to allow for the fact that the number density at the critical level was computed for travel vertically upwards whereas (5.7) includes molecules travelling in any upward direction. The integral in (5.7) may be evaluated by parts to give

$$\dot{N} = \frac{1}{8}n_H(z_c)\left(\frac{\alpha}{\pi}\right)^{1/2}\left(V_c^2 + \frac{1}{\alpha}\right)\exp(-\alpha V_c^2)$$ (5.8)

$$= \beta n_H(z_c), \text{ say}$$ (5.9)

where β has the dimensions of a velocity.

For hydrogen and $T = 1000\,\mathrm{K}$, $\beta \simeq 1.6\,\mathrm{m\,s^{-1}}$.

The flux $\dot{N}$ of hydrogen atoms has, of course, to be maintained right through the atmosphere up to the escape level. Above the turbopause, it is maintained predominantly by molecular diffusion, the net flux of atoms at any level due to diffusion being (omitting the rather small thermal diffusion term)

$$D\left\{\frac{dn_H}{dz}-\left(\frac{dn_H}{dz}\right)_e\right\} \tag{5.10}$$

where D is the diffusion coefficient and $(dn_H/dz)_e$ the vertical gradient of n_H under equilibrium conditions which from the hydrostatic equation is $-n_H/H_H$, H_H being the scale height for hydrogen atoms. Carrying through the calculation (problem 5.5) shows that to provide the required upward flux of hydrogen atoms, the ratio of hydrogen atom concentration at 120 km to that at the escape level must be substantially greater than it would be if no escape occurred.

5.4 The energy balance of the thermosphere

Above the turbopause at ~120 km altitude, we have already seen in §5.2 that, because of the low density and the positive temperature gradient with altitude, molecular diffusion is more important than eddy diffusion. The same is true for heat transfer. The dominant processes controlling the temperature of the thermosphere are the absorption of solar radiation in the extreme ultraviolet (wavelengths below 91 nm) by atomic oxygen, balanced by molecular conduction of heat. Of lesser importance under normal conditions are radiative transfer by far infrared transitions within the ground state of atomic oxygen and transfer of energy from particles which make up the *solar wind*. Under disturbed conditions the latter can become the dominant source of energy.

Ignoring, for the moment, variations through the day, consider equilibrium conditions for the atmospheric layer above level z. Since the conduction of heat across the top of the atmosphere is zero the solar radiation absorbed within a layer averaged over a day must be equal to the conduction of heat out of the bottom of the layer, i.e.

$$\varepsilon\{\bar{I}(\infty)-\bar{I}(z)\} = \lambda(z)\frac{dT(z)}{dz} \tag{5.11}$$

where $\bar{I}(z)$ is the average incident solar radiation, $\lambda(z)$ the coefficient of thermal conductivity and $T(z)$ the temperature, all at level z. The factor ε allows for the fact that not all the energy absorbed will be transformed into heat; some will be re-radiated out of the region altogether and some of the ionized or dissociated particles will diffuse out of the region to recombine elsewhere. According to simple kinetic theory, λ is proportional to the mean molecular velocity $\bar{c}$ which varies as $T^{1/2}$ but is independent of the pressure.

Knowing the distribution of atomic oxygen (§5.2) and its ultraviolet properties and also making some assumption about the value of ε, (5.11) may be solved numerically (problem 5.7). Fig. 5.3 shows the result of such calculations for different solar activities.

The problem of thermospheric heat balance may also be written in time-dependent form so that the diurnal variation may be estimated. Fig. 5.3 shows the variation of thermospheric temperature through the day as derived from such a calculation.

5.5   The photodissociation of oxygen

A fundamental photochemical process in the atmosphere is the dissociation of oxygen molecules through the absorption of ultraviolet radiation of frequency ν,

$$O_2 + h\nu \rightarrow O + O \qquad (5.12)$$

which occurs at wavelengths less than 246 nm. The resulting oxygen atoms are highly reactive; for instance, they are involved in the formation and destruction of ozone which will be addressed in the next section.

If at any level z, F_ν is the intensity of solar radiation incident per unit area, the rate of formation $D_\nu d\nu$ of oxygen atoms per unit volume through the absorption of solar radiation will be

$$D_\nu d\nu = -\beta_\nu d\nu \frac{dF}{d\nu} \qquad (5.13)$$

where β_ν is an efficiency factor relating the number of molecules dissociated to the amount of energy absorbed.

From (2.1) and (2.2) (see also (4.3)), (5.13) becomes

$$D_\nu = \beta_\nu k_\nu c\rho \sec\theta \exp\left\{ -\int_z^\infty k_\nu c\rho \sec\theta \, dz \right\} \qquad (5.14)$$

where θ is the angle of incidence of the solar radiation, k_ν the absorption coefficient of molecular oxygen at frequency ν, c the mixing ratio of molecular oxygen (substantially constant at the levels of interest) and ρ the atmospheric density at level z.

The dissociation rate in (5.14) can be written in terms of the pressure p rather than the height z by using the hydrostatic equation (1.2) and the equation of state (1.3) to give

Fig. 5.4. A Chapman layer. A plot of the dissociation rate of molecular oxygen from (5.15) against the height-like variable $-\ln(p/p_m)$.

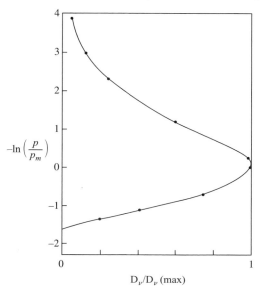

$$D_\nu = \frac{\beta_\nu g}{MRT}\left(\frac{K_\nu cp\sec\theta}{g}\right)\exp\left(\frac{-k_\nu cp\sec\theta}{g}\right) \tag{5.15}$$

The form of D_ν is plotted in fig. 5.4. Its maximum $D_\nu(max)$ is at p_m where

$$\frac{k_\nu cp_m\sec\theta}{g} = 1 \tag{5.16}$$

It is known as a *Chapman layer* after Sydney Chapman who first described it in 1930 in connection with the formation of layers of ionization in the upper atmosphere. We shall meet it again in §12.6, where we will find that thermal emission from the atmosphere to space in the infrared at a given frequency originates from a layer of similar character.

Because the absorption coefficient k_ν varies considerably over the absorption band responsible for oxygen dissociation (see appendix 9), when integration over the whole band is carried out, the height range of the layer involved becomes substantially greater.

5.6 Photochemical processes

We have already seen the importance of ozone in determining the temperature structure of the stratosphere and mesosphere. The most important photochemical processes in the atmosphere, therefore, are those which lead to the formation and destruction of ozone. The simple 'classical' theory of ozone due to Chapman (1930) considered reactions involving oxygen only, namely:

$$O_2 + hv \rightarrow O + O \qquad\qquad (J_2) \qquad\qquad (5.12)$$

$$O_3 + O + M \rightarrow O_2 + M \qquad\qquad (k_1) \qquad\qquad (5.17)$$

$$O_3 + O_2 + M \rightarrow O_3 + M \qquad\qquad (k_2) \qquad\qquad (5.18)$$

$$O + O_3 \rightarrow 2O_2 \qquad\qquad (k_3) \qquad\qquad (5.19)$$

$$O_3 + hv \rightarrow O_2 + O \qquad\qquad (J_3) \qquad\qquad (5.20)$$

The first of these reactions (5.12) has already been considered in §5.5.

Equations (5.12) and (5.17) to (5.20) describe photodissociation in the presence of solar radiation, the first occurring for wavelengths less than 246 nm and the second for wavelengths less than 1140 nm but most strongly below 310 nm. J_2 and J_3 are dissociation rates per molecule per second, which will, of course, vary with altitude. k_1, k_2 and k_3 are reaction rates defined such that for reactions between two species present in number densities n_1 and n_2 respectively per unit volume, the number of reactions per second will be the product $n_1 n_2$ multiplied by the reaction rate. Reactions (5.17) and (5.18) are three-body collisions, the third body M being required to satisfy energy and momentum conservation simultaneously. For these reactions to obtain the number of reactions per second, the appropriate reaction rate is multiplied by the product of the number densities of the three species involved.

In the stratosphere, reaction (5.17) is slow and may be neglected. Reactions (5.18) and (5.20), in which 'odd' oxygen particles (i.e. O and O_3) are converted into each other, are much faster than reactions (5.12) and (5.19) in which 'odd' oxygen particles are created or destroyed. Because of this, in the stratosphere, equilibrium between the O and O_3 concentration is governed by (from the law of Mass Action)

$$J_3 n_3 = k_2 n_2 n_1 n_M \qquad\qquad (5.21)$$

where n_1, n_2, n_3 and n_M are respectively the concentrations of O, O_2, O_3 and all molecules.

The balance between creation and destruction of odd oxygen particles is described by

$$2J_2 n_2 = 2k_3 n_1 n_3 \qquad\qquad (5.22)$$

From (5.21) and (5.22) the equilibrium ozone concentration is

Fig. 5.5. Values of dissociation rates J_2 and J_3 averaged over a day for typical mid-latitude conditions. (From Crutzen, 1971)

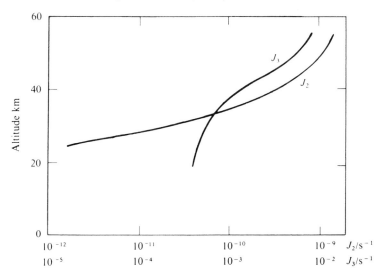

$$n_3 = n_2 \left(\frac{J_2 k_2 n_M}{J_3 k_3} \right)^{1/2} \tag{5.23}$$

The rates J_2 and J_3 depend differently on altitude (fig. 5.5), J_2 dropping off with decreasing altitude more rapidly than J_3, mainly because of the overlap of ozone and oxygen absorption in the 200 nm region. A peak of ozone concentration, therefore, occurs (fig. 5.6). Agreement between the computed ozone profile and that observed is not very good, because (1) other reactions involving minor constituents such as NO, NO_2 or other particles such as H, OH, HO_2 themselves formed photochemically are involved in the formation and destruction of O and O_3, and because (2) the lifetime of ozone molecules at levels below 30 km is long (problem 5.13) so that ozone is redistributed by atmospheric motions. We shall look briefly at these effects in turn.

Regarding the influence of minor constituents, processes removing 'odd' oxygen from the stratosphere can be visualized in terms of catalytic cycles in which the conversion of a species X to XO and back again destroys two 'odd' oxygen particles; i.e.

$$X + O_3 \rightarrow XO + O_2 \tag{5.24}$$

followed by

$$XO + O \rightarrow X + O_2 \tag{5.25}$$

Fig. 5.6. (a) Height distribution of ozone at 45° latitude, summer (A) as observed and (B) as calculated from 'classical' pure oxygen photochemical theory (after Dütsch, 1968). (b) Total ozone in a vertical column (expressed as the depth of the column of gas if isolated from the rest of the atmosphere and at STP in units of 10^{-5} m, sometimes known as Dobson units) plotted as a function of latitude and season (after Dütsch, 1971). Note the maxima in spring at high latitudes in both hemispheres which result from transport by atmospheric motions.

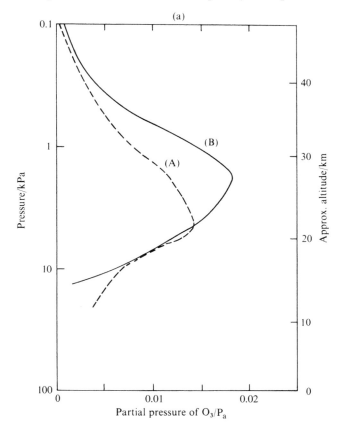

leads to the net result of

$$O + O_3 \rightarrow 2O_2 \tag{5.26}$$

Important species X involved in such catalytic cycles in the stratosphere are H, NO and Cl (problem 5.11).

The distribution in the atmosphere of constituents such as NO and Cl is influenced strongly by human activities. Anthropogenic emissions of NO result from the burning of fossil fuels especially from transport and from biomass burning (e.g. through deforestation). Free chlorine is released through photochemical action in the stratosphere on compounds

Fig. 5.6. (*continued*)

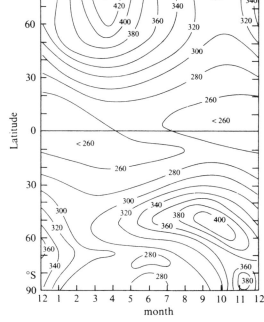

such as the chlorofluorocarbons (CFCs) widely used in refrigeration, aerosol sprays and industrial processes. Because of the substantial destruction of the ozone layer (for instance the presence of the 'ozone hole' in the stratosphere over Antarctica discovered in 1986) resulting from catalytic reactions involving these and other constituents the international community has taken action through the Montreal Protocol to phase out emissions of ozone depleting chemicals.

The detailed chemistry that occurs in the stratosphere is much more complex than the few simple reactions we have presented. The radical species we have mentioned react rapidly with each other as well as with ozone and under some circumstances ozone can be created rather than destroyed by adding appropriate pollutants (more details in World Meteorological Organization, 1994, and Wayne, 1991).

The second large influence on the ozone distribution mentioned above is that of atmospheric motion. Reference to fig. 5.6 will illustrate the effect. Although on photochemical theory it would be expected that the maximum ozone would occur in tropical regions where there is maximum solar radiation and a minimum in polar regions, in fact the reverse is the case. To account for this Brewer (1949) proposed what has become known as the *Brewer–Dobson* circulation in which air enters the stratosphere through upward motion in the tropics (cf. §10.9). Ozone-rich air is then transported polewards and downwards so that a large concentration of ozone occurs at high latitudes in the lower stratosphere where its photochemical lifetime is very long. Observations of the ozone mixing ratio provide an important tracer of these motions. During the winter a great deal of planetary wave activity is observed in the stratosphere (§8.6) the effect of which on transport processes needs to be taken into account for a complete picture of stratospheric behaviour to be realized (§10.9).

Some photochemical reactions give rise to excited molecular species. The resulting radiation is known as *airglow*. A particularly important reaction is

$$H + O_3 \rightarrow OH^* + O_2 \tag{5.27}$$

which occurs near the mesopause (~90 km) and leads to vibrationally excited OH molecules which radiate in the near infrared. The amount of emission is large and makes a significant contribution to the energy budget of the mesopause region (problem 5.14).

5.7　Breakdown of thermodynamic equilibrium

In the discussion of radiative transfer in chapters 2 and 4 the assumption of local thermodynamic equilibrium (LTE) enabled Kirchhoff's law to be applied so that the Planck black-body function could be employed as the source function in the equation of transfer (2.3). At high levels in the atmosphere LTE is no longer a good assumption, and the molecular processes involved need to be considered in more detail.

Consider a system (fig. 5.7) of the ground state and first excited state of a particular vibrational mode. At atmospheric temperatures very few molecules will be in states higher than the first and we shall neglect these higher states in this simplified treatment. Molecules can be excited into the upper level either by absorbing radiation of the appropriate frequency or through the effect of collisions. Molecules can lose their vibrational energy and transfer back to the ground state by emitting radiation either spontaneously or by stimulated emission or again through collisions. If the volume of gas under consideration is completely enclosed so that there is no net gain or loss by radiation a Boltzmann balance between the rate of excitation and de-excitation will be provided through the radiation and collision processes considered separately, and the ratio of the popula-

Fig. 5.7. A two-level system with Einstein coefficients: A_{21} for spontaneous emission, B_{12} for absorption, B_{21} for stimulated emission, a_{21} and b_{12} are probabilities per unit time of relaxation or excitation by collisional processes.

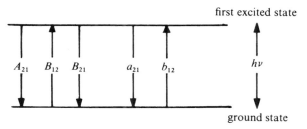

tion of the upper level to that of the lower level will be the Boltzmann factor $\exp(-h\nu/kT)$. If now part of the enclosure is removed so that some net radiation is gained or lost from the system a Boltzmann balance cannot be maintained through absorption and emission processes. However, it is still possible that the rates of excitation and de-excitation by collision will be sufficiently rapid that they dominate over the radiation processes; the population of the upper level will then be affected only a little by the radiative gain or loss. This is the situation of local thermodynamic equilibrium. We may expect it no longer to apply when (1) there is a significant net radiative gain or loss from the excited level in question and when (2) the rates of excitation or de-excitation by collisions are comparable with the rates of excitation or de-excitation by absorption or emission.

To set up the problem, the first treatment of which was by Milne (1930), we need to express parameters describing the radiation field such as the absorption coefficient and the source function in terms of the detailed molecular rate constants of fig. 5.7.

First considering emission, the radiation power emitted by the element of volume dv (fig. 5.8) containing ndv molecules, between frequencies ν and $\nu + d\nu$, and in solid angle $d\omega$ is $k_\nu n J_\nu ds dA d\nu d\omega$ where k_ν is the absorption cross-section per molecule – an expression which serves to define the source function J_ν. Assuming that J_ν is isotropic and, over the spectral region occupied by the absorption band, does not vary with ν, since $\int_{\Delta\nu} k_\nu d\nu = S$ (the total band strength of the vibrational transition, i.e. integrated over all rotational structure within the whole band of width $\Delta\nu$), the total power emitted by the element is $4\pi S n J_\nu d\nu$.

Turning this expression for total power emitted into a number of photons per unit time enables it to be related to the Einstein coefficient A_{21} (fig. 5.7), i.e.

Fig. 5.8

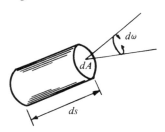

$$4\pi SnJ_v(hv)^{-1} = A_{21}n_2 \tag{5.28}$$

Note that in (5.28) we have neglected stimulated emission compared with spontaneous emission as for the bands of interest under atmospheric conditions spontaneous emission dominates.

In a similar way absorption of radiation in the element of volume dv can be considered. If I_v is the incident intensity within the solid angle $d\omega$ and the spectral interval dv, the power absorbed will be $k_vI_vndvdvd\omega$ which after integration gives, for the total power absorbed,

$$4\pi Snd v\bar{I}_v$$

where $\bar{I}_v = \dfrac{1}{4\pi S}\int_{\Delta v}\int_{4\pi} k_v I_v d\omega dv$ \hfill (5.29)

i.e. the intensity averaged with respect to ω and v. Now the rate of absorption of photons per unit volume is also equal to $B_{12}n_1\rho_v$ where ρ_v is the energy density of radiation at frequency v within the volume dv. *Since* $\rho_v = 4\pi\bar{I}_v c^{-1}$, *we therefore have*

$$4\pi Sn\bar{I}_v(hv)^{-1} = B_{12}n_1 4\pi c^{-1}\bar{I}_v \tag{5.30}$$

Referring again to fig. 5.7 equilibrium between the population of the levels requires

$$n_1(B_{12}4\pi\bar{I}_v c^{-1} + b_{12}) = n_2(A_{21} + a_{21}) \tag{5.31}$$

Substituting in (5.31) for n_1 and n_2 from (5.28) and (5.30) results in an expression for the source function

$$J_v = \frac{\bar{I}_v + (cb_{12}A_{21}/4\pi a_{21}B_{12})\phi}{1+\phi} \tag{5.32}$$

where $\phi = a_{21}/A_{21}$

In the limit where local thermodynamic equilibrium applies collisional activation and deactivation dominates so $\phi \to \infty$ and $J_v \to B_v$, the Planck function. The quantity in brackets in (5.32) must therefore be B_v and

$$J_\nu = \frac{\bar{I}_\nu + \phi B_\nu}{1 + \phi} \qquad (5.33)$$

Note that as $\phi \to 0$, $J_\nu \to \bar{I}_\nu$ and we have isotropic scattering.

If ψ is written for the heating rate, or the rate of increase in energy per unit volume due to radiation processes

$$\psi = -\int_{\Delta\nu} \int_{4\pi} \frac{dI_\nu}{ds} d\omega \, d\nu \qquad (5.34)$$

Substituting for dI/ds from the equation of transfer (2.3) (with J_ν written for B_ν):

$$\psi = -\int_{\Delta\nu} \int_{4\pi} k_\nu n (I_\nu - J_\nu) d\omega \, d\nu$$
$$= 4\pi S n (\bar{I}_\nu - J_\nu) \qquad (5.35)$$

from (5.29). Substituting for $\bar{I}_\nu$ from (5.33) we have

$$J_\nu - B_\nu = \frac{\psi}{4\pi S n \phi} \qquad (5.36)$$

Equation (5.36) expresses the difference between the source function and the black-body function in terms of the heating rate and the ratio ϕ of the probability of relaxation from the upper state through collisional processes a_{21} to that through radiation processes. As we should expect from the discussion at the beginning of this section, $J_\nu \simeq B_\nu$, i.e. LTE exists when the heating rate is small or when collisional processes dominate, i.e. when ϕ is large.

The most important application in the atmosphere of the theory of non-LTE radiative transfer is to emission by the $15\,\mu m$ vibration–rotation band of carbon dioxide, which for the LTE case was considered in §4.8. With the same approximation as was employed there, that cooling to space is the dominant term, for the non-LTE case (4.24) applies with B_ν replaced by J_ν i.e.

$$-\psi = \int_{\Delta\nu} \pi J_\nu \frac{d\tau_\nu^*(z,\infty)}{dz} dz \qquad (5.37)$$

with $\quad \tau_\nu^*(z,\infty) = \exp\left(-\int_z^\infty \frac{5}{3} k_\nu n \, dz'\right) \qquad (5.38)$

Over the spectral interval in which the band is contained, J_ν may be considered constant and

$$-\psi = \frac{5}{3} \pi J_\nu n S \overline{\tau_\nu^*} \qquad (5.39)$$

where the quantity

$$\overline{\tau_v^*} = \frac{1}{S} \int_{\Delta v} k_v \tau_v^*(z,\infty) dv \tag{5.40}$$

is the probability of a photon emitted by a carbon dioxide molecule at level z getting out to space. Substituting for J_v from (5.36) in (5.39) we have

$$-\psi = \frac{\frac{5}{3} \pi S n \overline{\tau_v^*} B_v}{1 + 5\overline{\tau_v^*}/12\phi} \tag{5.41}$$

Under conditions of LTE, the second term in the denominator of (5.41) is zero. The degree to which the cooling rate, therefore, differs from its LTE value depends not only on ϕ but also on $\overline{\tau_v^*}$ the transparency of the atmosphere to space.

For the $15\,\mu m$ carbon dioxide band, and collisions with air molecules, the collisional relaxation time $a_{21}^{-1} \approx 3 \times 10^{-5}\,s$ at standard pressure and $210\,K$ (an average temperature for the mesosphere). Since a_{21} is mainly dependent on the collision frequency it is approximately proportional to pressure. The radiative lifetime $A_{21}^{-1} = 0.74\,s$. We therefore have $\phi \approx 2.4 \times 10^4 p$ (where p is pressure in atm) and $\phi = 1$ at a pressure of $4\,Pa$ or a height of $73\,km$. Because, at this level $\overline{\tau_v^*}$ is small (problem 5.16), the LTE approximation applies to somewhat higher altitudes, in fact to about the $80\,km$ level. Above $80\,km$ the second term in the denominator becomes more and more significant. For instance at the $10^{-6}\,atm$ level ($\sim 96\,km$), the rate of cooling is only 0.05 of what it would be under LTE (problem 5.18 and fig. 5.9).

In the above discussion, the cooling to space approximation has been used. We conclude this section by writing down the heating rate equations in a particularly useful form for numerical computation in which the radiative transfer between all layers is included correctly.

Consider the atmosphere divided into a number of discrete layers. With the plane parallel approximation of §4.7 and for a frequency range sufficiently wide to include much or all of a vibration–rotation band, but sufficiently narrow that B_v or J_v is constant within the range, (2.4) and (4.23) can be written in finite difference form

$$\psi_k = \sum_j C_{kj} J_{vj} + \psi_{Sk} \tag{5.42}$$

where ψ_k is the heating rate at level k, J_{vj} is the source function at level j, C_{kj} is an element of a matrix known as the *Curtis matrix* which depends on atmospheric transmission and the summation is over all atmospheric

Fig. 5.9. Heating rates for the $15 \, \mu m$ band of carbon dioxide for a mean atmosphere, with collisional relaxation times at standard pressure (a) 2×10^{-6} s, (b) 1×10^{-5} s, (c) 3×10^{-5} s. (After Williams, 1971)

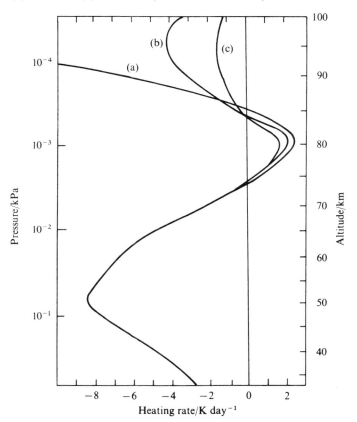

layers. Heating, due to solar absorption ψ_{Sk} has been included to make the treatment as general as possible. In the same notation, (5.36) is

$$J_k = B_k + E_k \psi_k \qquad (5.43)$$

where $E_k = (4\pi S n \phi)^{-1}$ appropriate to level k. In matrix notation, (5.42) and (5.43) are

$$\Psi = \mathbf{C}\mathbf{J} + \Psi_S$$

$$\mathbf{J} = \mathbf{B} + \mathbf{E}\Psi \qquad (5.44)$$

which may be solved to give

$$\Psi = (\mathbf{I} - \mathbf{C}\mathbf{E})^{-1}(\mathbf{C}\mathbf{B} + \Psi_S) \qquad (5.45)$$

$\mathbf{I}$ being the unit matrix.

In (5.45) $\mathbf{C}$ describes atmospheric transmission between different levels which for a uniformly mixed gas such as carbon dioxide may be computed once and for all for a variety of standard atmospheric conditions (the dependence of transmission on temperature is small). The vector $\mathbf{B}$ describes a particular atmospheric temperature profile. The form of (5.45), therefore, is particularly convenient for numerical computation of radiative heating rates for different atmospheric conditions.

Problems

5.1　Assuming different uniform temperatures above the turbopause at 120 km, estimate for different levels of solar activity the level above which the most abundant constituent is helium.

5.2　Show that a body of cross-sectional area A travelling at velocity V through a gaseous medium of density ρ experiences a drag force equal to $\frac{1}{2}\rho A c_D V^2$ where c_D is the drag coefficient. Hence estimate for standard atmospheric conditions the change in orbital period per orbit for a satellite of mass 100 kg with $A = 1\,\mathrm{m}^2$, $c_D = 2$, in a circular orbit at 600 km altitude. Observation of the details of satellite orbits is an important method of measuring density in the thermosphere.

5.3　Calculate the value of β (5.9) for helium atoms for $T = 1000\,\mathrm{K}$. Estimate the temperature which would be necessary for the value of β for helium to be $1.6\,\mathrm{m\,s}^{-1}$, i.e. equal to that for hydrogen at $T = 1000\,\mathrm{K}$.

5.4　Show that, from simple kinetic theory, the diffusion coefficient D of hydrogen atoms present in small proportion in air at temperature T and total number density n is proportional to $T^{1/2}n^{-1}$.

　　　Hence show that if a constant scale height H is assumed for the atmosphere as a whole above the turbopause where the total number density is n_0 then the density of hydrogen atoms is given by the differential equation

$$\frac{dn_H}{dz} + \frac{n_H}{H_H} = -\frac{\dot{N}n_0}{CT^{1/2}}\exp\left\{\frac{-(z-z_0)}{H}\right\} \tag{5.46}$$

where C is a constant.

5.5　Integrate (5.46). From your solution, given that n_H (120 km) $= 2 \times 10^{11}\,\mathrm{m}^{-3}$, $C = 2 \times 10^{20}\,\mathrm{m}^{-1}\,\mathrm{s}^{-1}\,\mathrm{K}^{-1/2}$ and finding suitable values for the other quantities from tables or diagrams, find the value of n_H at the escape level and hence the rate of escape of hydrogen atoms.

5.6 It has been suggested that oxygen present in the earth's atmosphere has resulted from the dissociation of water molecules in the high atmosphere followed by the escape of hydrogen. Assuming constant conditions calculate the loss of hydrogen during the earth's history ($\sim 4.5 \times 10^9$ years) and test the plausibility of this hypothesis. (This possible source is now believed to be small compared with oxygen produced through photosynthesis.)

5.7 Solve (5.11) numerically in the following way (you will need to write a simple computer programme). Assume a pure oxygen atmosphere above a lower boundary at $z = 120\,km$, $T = 300\,K$ and calculate from the hydrostatic equation the number density of atomic oxygen as a function of altitude for an isothermal atmosphere. Assume an absorption cross-section for atomic oxygen of $1.2 \times 10^{-21}\,m^2$, a value of the product εI_∞ of $1.1 \times 10^{-3}\,W\,m^{-2}$, a value of $\lambda = AT^{1/2}$ where $A = 3.6 \times 10^{-3}\,J\,m^{-1}\,s^{-1}\,K^{-3/2}$ and solve (5.11) for layers beginning with the lower boundary. From the resulting values of T, recalculate a new density profile and continue the iteration until a satisfactory solution has been obtained.

5.8 For the temperature profile of appendix 6 calculate the total potential energy (thermal and gravitational) above 120 km as given by (3.19). Compare this with the energy absorbed during an eight hour period (assuming the sun overhead). Hence make a crude estimate of the amplitude of diurnal variation in the exosphere. Compare your estimate with fig. 5.3.

5.9 Consider the maximum value of D_ν in (5.15) and derive (5.16). Find the value of p_m for the sun overhead and for $\nu = 210\,nm$ where the O_2 absorption cross-section per molecule is $10^{-23}\,cm^2$.

5.10 Assume an atmosphere at uniform temperature and show that the form of (5.15) in terms of height z is

$$\frac{D_\nu}{D_\nu(max)} = \exp\left\{1 - \left(\frac{z - z_m}{H}\right) - \exp\left[-\left(\frac{z - z_m}{H}\right)\right]\right\}$$

where z_m is the height where $D_\nu = D_\nu(max)$ and where H is the scale height.

5.11 Compare the rate of destruction of ozone at 40 km by the catalytic cycle of equations (5.24) and (5.25) with X = Cl with destruction through equation (5.19). Take the concentration of ClO to be 10^8 cm^{-3}, the rate constant of (5.24) with X = Cl to be $5 \times 10^{-11}\,m^3\,s^{-1}$ and other data as given in problem 5.13.

5.12 The Brewer–Dobson circulation mentioned in §5.16 was also introduced to account for the very low water vapour content of the
stratosphere. Supposing air always enters the stratosphere through
the tropical tropopause estimate from the tables in appendices 2
and 5 what value of water vapour mixing ratio can be expected in
the stratosphere.

5.13 Associated with the equilibrium of ozone as described in §5.6, there
are two very different time constants, the first being associated with
changes in the relative amount of O and O_3, the combined number
densities of O and O_3 being kept constant, and the second being
associated with changes in the sum of the amounts of O and O_3,
the ratio between them during the change being given by (5.21).
Write down differential equations for changes under these two
approximations and hence deduce expressions for the two time
constants. Calculate values for them at 40 km, 30 km and 20 km,
given that $k_2 = 10^{-33}\,\mathrm{cm}^6\,\mathrm{s}^{-1}$, $k_3 = 10^{-15}\,\mathrm{cm}^3\,\mathrm{s}^{-1}$. Take values of J_2, J_3 and
n_3 from figs. 5.5 and 5.6.

   Note that at the lower levels the second time constant is
very long indeed so that the ozone mixing ratio becomes a very
good 'tracer' of atmospheric motions.

5.14 Measurements of the OH airglow show that approximately 10^{12}
photons s^{-1} of wavelength between $1\,\mu\mathrm{m}$ and $3\,\mu\mathrm{m}$ are emitted
from a $1\,\mathrm{cm}^2$ column of atmosphere at levels near the meso-
pause. Assuming that the emission is uniformly distributed over
the region from 75 to 95 km altitude, estimate the cooling rate in
$\mathrm{K\,day}^{-1}$ due to this emission.

5.15 If W is the equivalent width of an absorption band for the path
between level z and space (absorber path length $u = |\int \frac{5}{3} c\rho\, dz|$),
show that $\overline{\tau_v^*}$ in (5.40) is given by

$$\overline{\tau_v^*} = \frac{1}{S}\frac{dW}{du}$$

where S is the strength of the band.

   Hence show that for a band of a gas present in constant
mixing ratio consisting of non-overlapping collision broadened
lines under either the 'weak' or 'strong' approximation the probability of a photon emitted from a level z getting out to space is
independent of height.

5.16 Calculate the value of $\overline{\tau_v^*}$ for a photon from the 15 μm carbon
      dioxide band under the non-overlapping 'strong' collision broad-
      ened approximation given that $\Sigma(S_i\gamma_{0i})^{1/2} = 1600\,\text{cm}^{-1}\,(\text{g cm}^{-2})^{-1/2}$ and
      $S = \Sigma S_i = 1.26 \times 10^5\,\text{cm}^{-1}\,(\text{g cm}^{-2})^{-1}$ and that the mass mixing ratio
      of carbon dioxide $= 5.5 \times 10^{-4}$.

5.17 Consider cooling from the 15 μm carbon dioxide band at the level
      where the pressure is 10^{-6} atm (~96 km). Here $\phi \ll 1$ and $\overline{\tau_v^*} \approx 1$.
      Substitute for B_v and for A_{21} ($A_{21} = 8\pi v^2 g_1 S/c^2 g_2$ where g_1 and g_2 are
      the statistical weights of the lower and upper levels respectively)
      in (5.41). Hence show that

$$-\psi = nhv\left\{\frac{g_2 a_{21}}{g_1[\exp(hv/kT)-1]}\right\}$$

      Notice that this result is independent of the band strength S, and
      that the quantity multiplying nhv is the probability per unit time of
      a quantum of energy being acquired on collision. Put in values
      appropriate to the 10^{-6} atm level and find the cooling rate in K per
      day.

5.18 Compare the value found in problem 5.17 with that which would
      occur if thermodynamic equilibrium prevailed.

5.19 Check the assumption $\overline{\tau_v^*} = 1$ for the lines in the 15 μm carbon
      dioxide band considered in problem 5.17 by finding the transmis-
      sion at the centre of a typical line of strength $s = 10^3\,\text{cm}^{-1}$
      $(\text{g cm}^{-2})^{-1}$ between the level 10^{-6} atm and space. (The absorption
      coefficient at the centre of a Doppler broadened line is $s\pi^{-1/2}\gamma_D^{-1}$ –
      see problem 4.8 for an expression for γ_D).

5.20 Calculate for carbon dioxide at STP (molecular diameter 4×10^{-10}
      m) the number of collisions per molecule per second from simple
      kinetic theory considerations. If the relaxation time for de-
      excitation of the v_2 vibration by collision is 6×10^{-6} s at STP, what
      is the probability of de-excitation on one collision?
           This very small probability is because hv for the v_2 vibra-
      tion is considerably greater than the average kinetic energy of a
      molecule ($\sim \frac{3}{2}kT$) at ordinary temperatures so that being able to
      transfer enough energy from translational motion to vibrational
      motion in any one collision is very unlikely.

6

Clouds

6.1   Cloud formation

It was demonstrated in §3.2 that ascent of damp air can lead to condensation and hence cloud formation. Four main kinds of clouds can be distinguished according to the type of ascending motions which produce them: (1) layer clouds formed by widespread, regular ascent such as occurs near the boundary between air masses having different characteristics (frontal zones) or because of orography; (2) layer clouds formed by vertical mixing (cf. problem 3.11); (3) convective clouds; (4) clouds formed under stable conditions by limited vertical motion as air passes over hills or mountains. The two first processes lead to *stratus* clouds or, when at higher levels, *cirrus* clouds, convective processes produce *cumulus* clouds, and associated with mountains are typically *lenticular* or *wave* clouds. Examples of these different cloud types will be found in the pictures taken from orbiting satellites in figs. 3.4, 3.5, 3.6 and 7.9.

Suitable nuclei on which condensation can occur are reasonably abundant in the atmosphere. Nuclei on which the freezing of a droplet may commence are, however, less abundant; the tops of clouds extending above the freezing level commonly contain supercooled droplets at temperatures well below 0°C.

The presence of clouds influences the energetics of the atmosphere in two main ways (fig. 6.1): (1) by the part which clouds play in the atmospheric water cycle; latent heat is released on condensation and liquid water is removed from the atmosphere on precipitation; (2) by scattering, absorption and emission of solar and terrestrial radiation clouds influence very strongly the atmosphere's radiation budget. In this chapter these two main influences will be discussed in turn.

Fig. 6.1. Schematic of the physical processes associated with clouds.

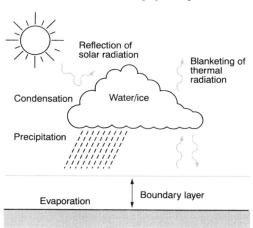

6.2   The growth of cloud particles

A cloud particle (droplet or ice crystal) has a typical diameter of
~10 μm whereas a rain drop of sufficient size to precipitate is ~1 mm in
diameter. Each precipitating particle, therefore, contains the same amount
of liquid water as about one million cloud particles. Two possible processes
are available for particle growth: (1) by diffusion of water vapour to the
cloud particles and subsequent condensation; (2) by collision and coalescence of particles.

First consider growth by condensation of an isolated spherical particle of mass m, radius r, density ρ_L in an environment where at distance x
from the particle the vapour density is ρ. The rate of particle growth is
sufficiently slow for a steady state diffusion equation to be applied to it
in which the mass of water vapour crossing any spherical surface in unit
time is independent of x and equal to $\dot{m}$, the rate of increase of mass of
the particle.

From Fick's law of diffusion we have, therefore,

$$\dot{m} = 4\pi x^2 D \frac{d\rho}{dx} \tag{6.1}$$

where D is the diffusion coefficient of water vapour in air. Integrating (6.1)
from the surface of the drop where the vapour density is ρ_r to a large distance away where it is ρ_∞ we have

$$4\pi D \int_{\rho_\infty}^{\rho_r} d\rho = \int_\infty^r \frac{\dot{m}}{x^2} dx \tag{6.2}$$

As water vapour condenses on the particles, latent heat is released at a rate $L\dot{m}$ where L is the latent heat of condensation. This has to be conducted away. A temperature gradient dT/dx is therefore set up, the equation describing the conduction being similar to (6.1), namely

$$L\dot{m} = -4\pi x^2 \lambda \frac{dT}{dx} \tag{6.3}$$

where λ is the thermal conductivity of air. Equation (6.3) may be integrated to give

$$L\dot{m} = 4\pi \lambda r (T_r - T_\infty) \tag{6.4}$$

where T_r and T_∞ are respectively the temperatures at the surface of the drop and at a large x.

From (6.2) and (6.4) the rate of droplet growth under different conditions can be evaluated (problems 6.1, 6.2 and 6.3). The process is particularly important at temperatures $\sim -10°C$ when a few ice crystals are present in a cloud which predominantly contains supercooled water drops (cf. fig. 3.6). The saturation vapour pressure over the water drops (see appendix 2) is about 10% more than that over the ice crystals which, therefore, grow at the expense of the water drops.

The other process of droplet growth is by collision and coalescence for which the rate of growth of a larger droplet of fall speed V in a population of smaller drops of fall speed v will be

$$\frac{dr}{dt} = \frac{\varepsilon w}{4\rho_L}(V - v) \tag{6.5}$$

where w is the mass of liquid water in the form of smaller droplets per unit volume of the cloud and ε is the collision efficiency. Since small droplets which approach each other tend to follow the streamlines of the air, they are unlikely to collide unless they are both large enough and unless there is sufficient size differential between them. Cloud particles grow to $30\,\mu m$ in diameter largely by condensation processes after which the collision efficiency becomes larger, subsequent growth continuing by coalescence (fig. 6.2).

6.3   The radiative properties of clouds

In chapter 2 when considering the simple radiative equilibrium model the assumption was made that solar radiation incident on the atmosphere is unmodified by the atmosphere and all absorbed at the surface; although the absorption of a small proportion of the solar radiation by ozone in the ultraviolet and by water vapour in the infrared were men-

Fig. 6.2. Rate of growth of cloud droplets for cloud with liquid water content $1\,\mathrm{g\,m^{-3}}$ (a) by coalescence assuming that distribution function of cloud droplet radii r given by $4r^2 r_m^{-3} \exp(-2r/r_m)$ with $r_m = 2\,\mu\mathrm{m}$, (b), (c), (d) by condensation assuming water vapour pressure is above saturation by (b) $4\,\mathrm{Pa}$, (c) $1.2\,\mathrm{Pa}$, (d) $0.25\,\mathrm{Pa}$: after Shishkin (see Matveev, 1967, 513).

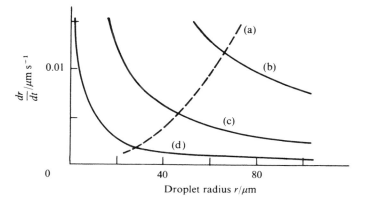

Fig. 6.3

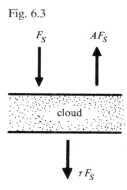

tioned (cf. fig. 2.1). Solar radiation is also modified on its passage through the atmosphere by scattering processes (§4.1). Clouds are also of great importance; because on average they cover ~50% of the earth's surface their influence on both the incoming and outgoing radiation is large.

Considering first the modification of solar radiation by clouds (fig. 6.3), if F_S is the incident solar flux on the top of the cloud, AF_S the reflected upward flux at the top of the cloud (A being the *cloud albedo*) and τF_S the downward flux at the bottom of the cloud (τ being its *transmissivity*), the amount of radiant energy absorbed is $F_S(1 - A - \tau)$. For a typical stratus cloud and for radiation integrated over the whole spectrum, typical values are $A = 0.5$, $\tau = 0.3$. Calculations of how A and τ depend on cloud thickness and on the properties of the cloud particles are presented in §6.4.

Because liquid water and ice absorb strongly throughout the infrared region, clouds are strong absorbers of terrestrial radiation. For considerations of the atmosphere's radiation budget it is satisfactory to assume that all clouds except cirrus absorb and emit terrestrial radiation as black bodies. Cirrus clouds are often thin and tenuous; although their albedo is small their transmissivity and emissivity may vary very considerably.

6.4   Radiative transfer in clouds

In chapters 2 and 4 radiative transfer theory was developed considering absorption and emission only. A simple extension of the theory enables scattering also to be taken into account.

Considering scattering particles distributed uniformly through a volume, a scattering coefficient σ can be defined in the same way as the absorption coefficient (2.1) so that the element of optical depth $d\chi = (k + \sigma)\rho \, dz$.

The absorption coefficient k takes into account absorption both by the particles and by the gas in between them. The coefficient σ accounts for scattering in all directions. A quantity called the *albedo for single scattering* ω is the ratio $\sigma/(k + \sigma)$.

As in chapters 2 and 4 we are considering transfer in the vertical dimension only (integrating over other angular dependences) with fluxes $F^\uparrow$ and $F^\downarrow$. The transfer equations become

$$\frac{dF^\downarrow}{d\chi^*} + F^\downarrow = \pi B(1 - \omega) + \omega\left\{ fF^\downarrow + (1 - f)F^\uparrow \right\} \tag{6.6}$$

$$-\frac{dF^\uparrow}{d\chi^*} + F^\uparrow = \pi B(1 - \omega) + \omega\left\{ (1 - f)F^\downarrow + fF^\uparrow \right\} \tag{6.7}$$

The first term on the right hand side is the emission term depending on the black-body function B in exactly the same way as in (2.3). The second term is the contribution to radiation transfer from the scattering processes. The fraction f is the proportion of the scattered radiation which in one scattering event goes into a forward direction while $(1 - f)$ is that which is scattered in a backward direction. For isotropic scattering $f = \frac{1}{2}$.

Equations (6.6) and (6.7) account properly for absorption, emission and for multiple scattering in cases where the restriction to one dimension is a satisfactory approximation. Their solution in the general case can be complicated. Here we shall treat only the case for solar radiation incident vertically downwards on the top of a cloud; the emission term (the first on the right hand side) in (6.6) and (6.7) will therefore be omitted.

It is instructive first to let $\omega = 1$ when we have pure scattering. Equations (6.6) and (6.7) may then be written

$$\frac{dF^\downarrow}{d\chi*} + (F^\downarrow - F^\uparrow)(1-f) = 0$$

$$\frac{dF^\uparrow}{d\chi*} + (F^\downarrow - F^\uparrow)(1-f) = 0 \qquad (6.8)$$

giving immediately

$$F^\downarrow - F^\uparrow = \text{constant}$$

and for a uniform cloud of optical thickness χ_0^* the boundary conditions are (fig. 6.3) $F^\downarrow(\chi* = 0) = F_s$, $F^\uparrow(\chi* = \chi_0^*) = 0$ so that the cloud albedo A and transmissivity τ respectively become

$$A = \frac{\chi_0^*(1-f)}{1 + \chi_0^*(1-f)} \qquad (6.9)$$

$$\tau = \frac{1}{1 + \chi_0^*(1-f)} \qquad (6.10)$$

Note that $A + \tau = 1$ as is required by conservation of energy.

When $\omega \neq 1$ (6.6) and (6.7) may be solved by differentiating again (6.6) and then substituting for $dF^\uparrow/d\chi*$ from (6.7). The solution is

$$F^\downarrow = C \exp(-\alpha\chi*) + D \exp\left[-\alpha(\chi_0^* - \chi*)\right]$$

with a similar expression for $F^\uparrow$ where

$$\alpha^2 = (1-\omega)(1+\omega-2\omega f) \qquad (6.11)$$

and C and D are constants. Applying the boundary conditions leads to

$$A = \beta \frac{1 - \exp\left(-2\alpha\chi_0^*\right)}{1 - \beta^2 \exp\left(-2\alpha\chi_0^*\right)} \qquad (6.12)$$

$$\tau = \frac{(1-\beta^2)\exp\left(-\alpha\chi_0^*\right)}{1 - \beta^2 \exp\left(-2\alpha\chi_0^*\right)} \qquad (6.13)$$

where $\beta = \frac{\alpha - 1 + \omega}{\alpha + 1 - \omega} \qquad (6.14)$

Note that as χ_0^* becomes large

$A \rightarrow \beta$ and $\tau \rightarrow 0$

The quantity β, therefore, is the albedo of a very thick cloud. For $\omega = 0.9997$, $f = 0.9$, values appropriate at a wavelength of ~0.7 μm to a stratus cloud with drops ~10 μm diameter, the value of β is 0.925. The optical thickness of the cloud χ_0^* needs to be large for its albedo to approach this value of β; for $\chi_0^* \simeq 125$ for instance, the albedo of this cloud $A = 0.90$. For a typical stratus cloud 0.5 km thick, $\chi_0^* \simeq 20$, giving $A \simeq 0.66$ (problem 6.9). Note in figs. 3.6(a) and 7.9 the intense brightness of the thickest clouds.

6.5   Cloud radiation feedback

In section 6.3 (see also fig. 6.1) it was mentioned that clouds interfere with the transfer of radiation in the atmosphere in two ways. Firstly, they reflect a certain proportion of solar radiation back to space, so reducing the total energy available to the system. Secondly, they act as blankets to thermal radiation from the earth's surface in a similar way to greenhouse gases. By absorbing thermal radiation emitted by the earth's surface below, and by themselves emitting thermal radiation, they act to reduce the heat loss to space from the surface.

Which effect dominates for any particular cloud depends on the cloud temperature and hence on the cloud height and on those properties (often known as optical properties) which determine the cloud's reflectivity to solar radiation and its interaction with thermal radiation. As shown in the last two sections, the latter depend on whether the cloud is of water or ice, on its liquid or solid water content (which depends also on how thick or thin it is) and on the average size of the cloud particles. In general for low clouds the reflectivity effect wins so they tend to cool the earth-atmosphere system; for high clouds, by contrast, the blanketing effect is dominant and they tend to warm the system. The overall feedback effect of clouds on the surface temperature, therefore, can be either positive or negative. Fig. 6.4 and problem 6.15 illustrate this.

A concept helpful in distinguishing between the two effects of clouds we have mentioned is that of *cloud radiative forcing*. Consider the net radiation (both solar and terrestrial components) leaving the top of the atmosphere above a cloud; suppose it has a value R. Now imagine the cloud to be removed, leaving everything else the same; suppose the net radiation leaving the top of the atmosphere is now R'. The difference $R' - R$ is the cloud radiative forcing. It can be separated into solar radiation and terrestrial radiation components. Values of the cloud radiative forcing deduced from satellite observations and as simulated in models are illustrated in fig. 6.5. On average it is found that the overall effect of clouds is

Fig. 6.4. Equilibrium surface temperature distributions in K calculated from an atmospheric radiative transfer model as a function of cloud amount for three cloud layers and for July conditions at 35N. The clouds are assumed to possess liquid water amounts in a vertical path of 140, 140 and 20 g m^{-2} and to occupy layers 91–85, 63–55 and 38–30 kPa respectively. (After Stephens & Webster, 1981)

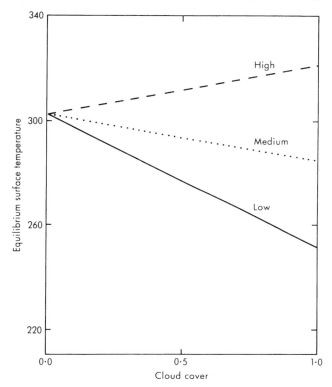

slightly to cool the earth-atmosphere system. The concept of cloud radiation forcing and the effect of clouds on climate are addressed further in section 14.7.

Problems

(See the appendices for some of the physical data necessary for the solution of problems.)

6.1 Integrate the Clausius–Clapeyron equation (3.13) for water vapour between temperatures T_r and $T_\infty (T_r - T_\infty \ll T_\infty)$ to give an expression in the following form for the corresponding saturation vapour pressures $p_s(T_r)$ and $p_s(T_\infty)$

Fig. 6.5. The cloud radiative forcing is made up of a solar radiation component and a thermal radiation component which generally act in opposite senses, each typically of magnitude between 50 and 100 W m^{-2}. The average net forcing is shown here for the period January to July as a function of latitude as observed from satellites in the two years of the Earth Radiation Budget Experiment (ERBE) and also as simulated by a climate model with different schemes of cloud formulation (RH, a simple threshold relative humidity scheme; CW, a scheme that includes cloud water as a separate variable and distinguishes between water and ice clouds; CWRP, as CW but with cloud radiative properties dependent on cloud water content). There is encouraging agreement between the models' results and the observations, but also differences that need to be understood. It is through comparisons of this kind that further elucidation of cloud radiation feedback will be achieved. (From UK Meteorological Office)

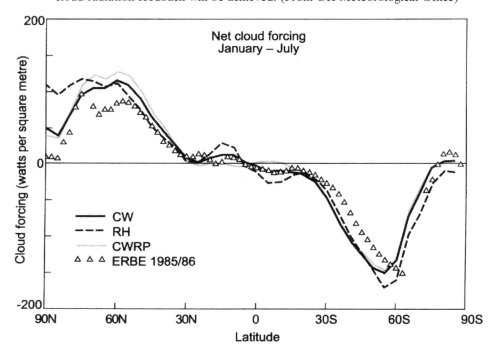

$$\ln\left[\frac{p_s(T_r)}{p_s(T_\infty)}\right] \simeq \frac{LM_r(T_r - T_\infty)}{RT_\infty^2} \tag{6.15}$$

(R is the gas constant and M_r molecular weight of water.)

6.2   Write (6.2) in terms of vapour pressure rather than vapour density. Using (6.4) and (6.15) derive the following expression for the rate of growth of a water droplet.

$$r\frac{dr}{dt} \simeq \frac{S-1}{\left[(L^2 M_r \rho_L / \lambda R T^2) + \{\rho_L R T / \rho_s(T_\infty) D M_r\}\right]} \tag{6.16}$$

where $S = p_\infty/p_s(T_\infty)$, p_∞ being the actual vapour pressure far away from the droplet. $S - 1$ ($\ll 1$) is the *supersaturation* of the vapour phase.

The effective degree of supersaturation depends also on (1) the droplet radius especially for drops $<1\,\mu$m in radius over which the equilibrium vapour pressure is considerably higher than over a plane surface, and (2) the purity of the water, the equilibrium vapour pressure being less for salt solutions. Because of these considerations very small drops only persist on hygroscopic nuclei.

6.3   For pure water use (6.16) to calculate the time taken for a droplet to grow from $2\,\mu$m radius (1) to $10\,\mu$m radius and (2) to $40\,\mu$m radius by condensation if the supersaturation is 0.05% (i.e. $S - 1 = 5 \times 10^{-4}$) and $T = 273\,$K. ($D = 0.23\,\text{cm}^2\text{s}^{-1}$.)

6.4   From Stokes' law for the viscous force F acting on a sphere of radius r falling with terminal velocity v in a fluid of viscosity η namely

$F = 6\pi\eta v r$

calculate the time taken for drops of radius $1\,\mu$m, $10\,\mu$m, $100\,\mu$m, to fall through $1\,$km. (η for air $= 1.7 \times 10^{-5}\,\text{kg}\,\text{m}^{-1}\text{s}^{-1}$.)

6.5   From values of the saturation vapour pressure over plane liquid water surfaces and plane ice surfaces given in appendix 2 for $-10°$C, calculate the time taken for an ice crystal in a cloud of water droplets at $-10°$C to grow from $1\,\mu$m radius to $100\,\mu$m radius, assuming that the crystal remains spherical.

Calculate also the temperature of the ice crystal. (Latent heat of sublimation of ice $= 2800\,\text{J}\,\text{g}^{-1}$, $D = 0.23\,\text{cm}^2\text{s}^{-1}$.)

6.6   For $\omega = 1$, find A and τ for $\chi_0^* = 20$ and $f = 0.9$. Compare with the values quoted in the chapter when $\omega = 0.9997$.

6.7   From (6.11) and (6.14) show that

$$\beta^2 = \frac{1 - \omega f - \alpha}{1 - \omega f + \alpha}$$

6.8   Derive (6.12) and (6.13).

6.9   For stratus cloud described in text with $\chi_0^* = 20$, find transmissivity τ. Also find the absorptivity of the cloud (i.e. $1 - \tau - A$).

6.10 The scattering cross-section at wavelength λ of a drop of radius r when the ratio $2\pi r/\lambda \gg 1$ is approximately $2\pi r^2$. Find the optical thickness of a cloud 0.5 km thick containing 150 drops cm^{-3} having $r = 5\,\mu$m. Remember $\chi_0^* \simeq 1.66\chi_0$.

6.11   At 2.3 μm, for drops having $r = 7\,\mu$m, $\omega = 0.988$, $f = 0.9$, find β and find the value of χ_0^* to give an albedo of 0.98β. Compare this value of χ_0^* with that for typical stratus.

6.12   At 10.6 μm, the wavelength of the CO_2 laser, $\omega = 0.36$, $f = 0.9$. Find β and the transmissivity of diffuse radiation at this wavelength in 100 m of fog having the same properties as the stratus cloud in the text. (Assume χ_0^* is the same at 10.6 μm as at 0.7 μm.)

6.13   In the visible part of the spectrum the albedo of the Venus clouds is approximately 0.9. Supposing the clouds to be very deep and that scattering is isotropic, what is the minimum value of ω for the cloud particles?

6.14   Why does the sky appear blue while clouds in general appear white?

6.15   For a simplified discussion of the cloud radiation feedback problem, consider a black surface at temperature T_0 illuminated by average solar radiation F such that $F = \sigma T_0^4$. Introduce an idealized cloud in front of the surface such that it is perfectly absorbing and emitting in the infrared but absorbs no solar radiation. The cloud's albedo for solar radiation is A. Consider the radiation balance of the cloud and the new radiation balance of the surface and solve for the cloud temperature T_c and the new surface temperature T_1 in terms of A and T_0. Show that if $A = 0.5$, $T_1 = T_0$ and that the behaviour of $T_1 - T_0$ for different cloud types agrees qualitatively with fig. 6.4. The reason for poor quantitative agreement is that in the simplified calculation the presence of the atmosphere has been ignored.

7

Dynamics

7.1   The material derivative

So far in making simple atmospheric models we have considered energy transfer by radiation, conduction and convection. In chapter 3 the atmosphere's radiative sources and sinks were shown to be driving a thermodynamic engine that generates kinetic energy as well as potential energy. Thermodynamic considerations alone cannot tell us much more about the form of this kinetic energy. In this chapter and those that follow, therefore, we turn our attention to the atmosphere's motions, our aim being to find simple descriptions or physical models which can aid our understanding of some of the main features of the atmosphere's circulation.

When applying the laws of motion to an element of fluid, it is important to realize that the fluid being considered is in general moving with respect to our chosen frame of reference. The fluid flow can be described or measured in two ways. One obvious way is to describe the flow at fixed points in space (or for the atmosphere, fixed with respect to the earth); this is called an *Eulerian* description. An anemometer or a wind vane, for instance, makes measurements at fixed points in space. Another way is to imagine moving with the fluid, describing the forces and their effect on a tiny parcel of fluid; this is called a *Lagrangian* description.

We shall frequently need to relate the rate of change of some quantity (say a scalar quantity ψ) following the fluid – this is called the *material* derivative and is denoted by $D\psi/Dt$ – to the *local* rate of change at a fixed point in the chosen frame of reference $\partial\psi/\partial t$.

Since ψ is varying in both space and time

$$D\psi = \frac{\partial \psi}{\partial t}dt + \frac{\partial \psi}{\partial x}dx + \frac{\partial \psi}{\partial y}dy + \frac{\partial \psi}{\partial z}dz$$

i.e. $$\frac{D\psi}{Dt} = \frac{\partial \psi}{\partial t} + u\frac{\partial \psi}{\partial x} + v\frac{\partial \psi}{\partial y} + w\frac{\partial \psi}{\partial z}$$

or $$\frac{D\psi}{Dt} = \frac{\partial \psi}{\partial t} + \mathbf{V}. \text{ grad } \psi \tag{7.1}$$

where $$u = \frac{dx}{dt}, \quad v = \frac{dy}{dt}, \quad w = \frac{dz}{dt}$$

are the components of the fluid's velocity $\mathbf{V}$ with respect to Cartesian axes in our frame of reference.

Equation (7.1) will be used constantly in the paragraphs that follow.

7.2   Equations of motion

Newton's second law of motion applied to an element of fluid of density ρ moving with velocity $\mathbf{V}$ in the presence of a pressure gradient ∇p and a gravitational field (described by $\mathbf{g}'$) is given by the equation, known as the *Navier–Stokes* equation (for its derivation see, for instance, Batchelor, 1967 or Andrews, 1999).

$$\frac{D\mathbf{V}}{Dt} = \mathbf{g}' - \frac{1}{\rho}\nabla p + \mathbf{F} \tag{7.2}$$

where $\mathbf{F}$ is the frictional force on the element which will be discussed more fully in chapter 9.

Equation (7.2) applies to an absolute or inertial frame of reference (i.e. say fixed with respect to the solar system). Our interest is in motion relative to axes fixed with respect to the earth's surface, which is rotating with angular velocity $\mathbf{\Omega}$. We therefore need to find an expression for DV/Dt appropriate to the rotating frame.

A vector $\mathbf{A}$ in frame Σ which is rotating at angular velocity $\mathbf{\Omega}$ with respect to frame Σ' will have a component of motion $\mathbf{\Omega} \wedge \mathbf{A}$ in frame Σ' due to the relative motion of the two frames (fig. 7.1) so that

$$\left(\frac{d\mathbf{A}}{dt}\right)_{\Sigma'} = \left(\frac{d\mathbf{A}}{dt}\right)_{\Sigma} + \mathbf{\Omega} \wedge \mathbf{A} \tag{7.3}$$

Repeating the procedure of equation (7.3) to the vector $d\mathbf{A}/dt$ we obtain for the second derivative

Fig. 7.1

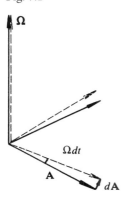

$$\left(\frac{d^2\mathbf{A}}{dt^2}\right)_{\Sigma'} = \left(\frac{d^2\mathbf{A}}{dt^2}\right)_{\Sigma} + 2\mathbf{\Omega} \wedge \left(\frac{d\mathbf{A}}{dt}\right)_{\Sigma} + \mathbf{\Omega} \wedge (\mathbf{\Omega} \wedge \mathbf{A})$$

In particular, if $\mathbf{A} = \mathbf{r}$, the position vector, we can obtain an expression for $D\mathbf{V}/Dt$ in the rotating frame,

$$\left(\frac{D\mathbf{V}}{Dt}\right)_{\Sigma'} = \left(\frac{D\mathbf{V}}{Dt}\right)_{\Sigma} + 2\mathbf{\Omega} \wedge \mathbf{V} + \mathbf{\Omega} \wedge (\mathbf{\Omega} \wedge \mathbf{r})$$

so that in the rotating frame (7.2) becomes

$$\frac{D\mathbf{V}}{Dt} + 2\mathbf{\Omega} \wedge \mathbf{V} + \mathbf{\Omega} \wedge (\mathbf{\Omega} \wedge \mathbf{r}) = -\frac{1}{\rho}\mathbf{\nabla}p + \mathbf{g}' + \mathbf{F}$$

or $$\frac{D\mathbf{V}}{Dt} + 2\mathbf{V} \wedge \mathbf{\Omega} - \frac{1}{\rho}\mathbf{\nabla}p + \mathbf{g} + \mathbf{F} \tag{7.4}$$

where $\mathbf{g} = \mathbf{g}' - \mathbf{\Omega} \wedge (\mathbf{\Omega} \wedge \mathbf{r})$ (7.5)

is the acceleration due to gravity and includes as it should the centrifugal term $\mathbf{\Omega} \wedge (\mathbf{\Omega} \wedge \mathbf{r})$ (problem 7.1).

The first term on the right hand side of (7.4) is the Coriolis term which applies particularly to moving particles in rotating frames. It is perpendicular both to the direction of motion and the earth's axis of rotation. As we shall see in this and the following chapters, the Coriolis term has a profound influence on atmospheric motion.

A convenient set of axes at any point on the earth's surface (fig. 7.2) has x directed towards the east, y towards the north and z vertically upwards (or more precisely in the direction of the vector defined by (7.5)). This set is not strictly Cartesian because the directions of the axes are

Fig. 7.2

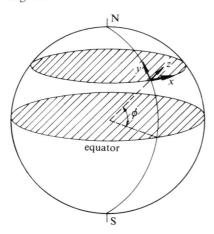

functions of position on the spherical earth. If u, v, w are the components of the velocity $\mathbf{V}$ in the x, y, z directions respectively and $\mathbf{i}$, $\mathbf{j}$, $\mathbf{k}$ are unit vectors directed along each axis then taking into account the way the axes change with position (problem 7.2)

$$\frac{D\mathbf{V}}{Dt} = \left(\frac{Du}{Dt} - \frac{uv\tan\phi}{a} + \frac{uw}{a}\right)\mathbf{i}$$

$$+ \left(\frac{Du}{Dt} + \frac{u^2\tan\phi}{a} + \frac{wv}{a}\right)\mathbf{j} + \left(\frac{Dw}{Dt} - \frac{u^2 + v^2}{a}\right)\mathbf{k} \qquad (7.6)$$

where ϕ is the latitude and a the earth's radius. Also

$$2\mathbf{V} \wedge \mathbf{\Omega} = 2\Omega(v\sin\phi - w\cos\phi)\mathbf{i}$$

$$- 2\Omega u\sin\phi\,\mathbf{j}$$

$$+ 2\Omega u\cos\phi\,\mathbf{k} \qquad (7.7)$$

The magnitudes of the various terms in (7.6) and (7.7) will be very different depending on the scale of the motion under study. In this chapter we shall be concerned with motions on what is generally known as the synoptic scale or larger, that is systems of typically 1000 km in horizontal dimension, very much larger than their vertical scale (of order 1 scale height or ~10 km). For this scale observed vertical velocities (typically $1\,\mathrm{cm\,s^{-1}}$) are very much smaller than horizontal velocities (typically $10\,\mathrm{m\,s^{-1}}$) so that, in the momentum equation (7.4) terms involving w can, to a first approximation, be neglected. Such motion is described as *quasi-horizontal*. Because of this, and because those terms in (7.6) which involve

a in the denominator are smaller by about one order of magnitude than the other terms (problem 7.3) and may therefore again to a first approximation be neglected, the equation of motion (7.4) may be simplified to become

$$\frac{D\mathbf{V}}{Dt} = f\mathbf{V} \wedge \mathbf{k} - \frac{1}{\rho}\nabla p + \mathbf{F} \qquad (7.8)$$

where $f = 2\Omega \sin \phi$ $\qquad (7.9)$

and where now

$$\frac{D\mathbf{V}}{Dt} = \mathbf{i}\frac{Du}{Dt} + \mathbf{j}\frac{Dv}{Dt}$$

and all the terms are vectors in the horizontal plane.

For the vertical direction, under the same approximation (problem 7.3) the hydrostatic equation (1.4) applies.

We now consider two important approximations to (7.8) in which for different situations some of the terms may be neglected in comparison to the others. Other approximations are mentioned in the problems at the end of the chapter.

7.3   The geostrophic approximation

For large scale motion away from the surface the friction $\mathbf{F}$ is small. Further, for steady flow with small curvature $D\mathbf{V}/Dt \simeq 0$. The resulting motion is known as *geostrophic*. The geostrophic velocity $\mathbf{V}_g$ is given by (fig. 7.3)

$$f\mathbf{V}_g \wedge \mathbf{k} = \frac{1}{\rho}\nabla p \qquad (7.10)$$

describing the familiar situation in which the wind blows parallel to the isobars, for the northern hemisphere in a clockwise direction around centres of high pressure (*anticyclones*) and anticlockwise around centres

Fig. 7.3. The geostrophic approximation.

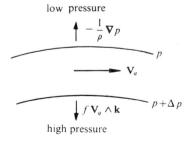

Fig. 7.4

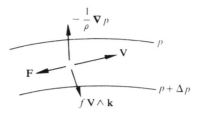

of low pressure (*depressions* or *cyclones*). For $\phi = 30°$, $\sin \phi = \frac{1}{2}$, $f = 7.29 \times 10^{-5}$ and for a pressure gradient of $0.2\,\text{kPa}\ (100\,\text{km})^{-1}$, the appropriate velocity for balance is $23.8\,\text{m}\,\text{s}^{-1}$.

The geostrophic approximation works well at heights above ~1 km (at lower levels friction becomes important) and for latitudes >~10°, so that under these conditions the wind velocity and direction may be deduced from a chart of isobars. Note that for horizontal motion, the Coriolis term is zero at the equator.

At lower levels where friction is not negligible (cf. chapter 9), since friction acts in a direction approximately opposite to **V**, the velocity can no longer be precisely parallel to the isobars (fig. 7.4). For a high pressure centre the balancing velocity has a component outwards from the centre and for a low pressure centre an inward component. To conserve the mass of air within the system, air therefore sinks over high pressure systems and rises over low pressure ones (§9.5), producing an important modification to the vertical velocities which would occur in the absence of friction.

7.4   Cyclostrophic motion

At low latitudes f is small and for motion having a large curvature as for instance a tropical cyclone, another approximation to (7.8) is the *cyclostrophic* one in which the acceleration of the air towards the centre (distance r away with the centre considered as the origin of co-ordinates) is balanced by the pressure gradient $\partial p/\partial r$

i.e. $$\frac{V^2}{r} = \frac{1}{\rho}\frac{\partial p}{\partial r} \qquad\qquad (7.11)$$

For a pressure gradient of $3\,\text{kPa}\ (100\,\text{km})^{-1}$ and $r = 100\,\text{km}$, $V \simeq 50\,\text{m}\,\text{s}^{-1}$ – typical values for a tropical cyclone. Note that the sense of the motion can be cyclonic or anticyclonic, but the air always blows around a region of low pressure.

If both the acceleration term and the Coriolis terms are included, the resulting solution is known as the *gradient wind* (problem 7.9).

The ratio of the acceleration term dV/dt to the Coriolis term is a dimensionless number known as the *Rossby number Ro*. For motion with speed V on a scale of typical horizontal dimension L,

$$Ro \simeq \frac{V^2/L}{fV} = \frac{V}{fL} \tag{7.12}$$

In mid-latitudes a typical value of Ro is 0.1 (problem 7.7) in which case the error introduced by the geostrophic approximation is ~10% (problem 7.8). The smallness of the Rossby number therefore is a measure of the validity of the geostrophic approximation.

7.5   Surfaces of constant pressure

If z is the height of a surface of constant pressure p on which two neighbouring points A and B have the same y co-ordinate

$$\frac{\partial p}{\partial x}\Delta x + \frac{\partial p}{\partial z}\Delta z = \Delta p = 0$$

Since $\quad \dfrac{\partial p}{\partial z} = -g\rho$ $\qquad\qquad\qquad\qquad\qquad\qquad$ (7.12)

we have

$$g\frac{\partial z}{\partial x} = \frac{1}{\rho}\frac{\partial p}{\partial x}$$

and similarly for

$$\partial z/\partial y$$

so that

$$g\rho\mathbf{\nabla}_p z = \mathbf{\nabla}_z p \tag{7.13}$$

where $\mathbf{\nabla}_p z$ denotes the gradient of z in the surface of constant pressure. Equation (7.13) relates the gradient of pressure in a horizontal surface to the gradient of height in a nearby surface of constant pressure.

Because of the variation of g with altitude and with latitude (problem 7.11), (7.13) is not easy to use as it stands. It is convenient to define geopotential Φ as

$$\Phi = \int_0^z g\,dz$$

If now the geostrophic approximation (7.10) is written in terms of Φ, we have

$$\mathbf{V}_g = \frac{1}{f}\mathbf{k}\wedge\nabla_p\Phi \tag{7.14}$$

Notice that the density ρ has disappeared from the equation; it is therefore applicable without change to any level in the atmosphere. For this reason standard meteorological charts are plotted as contours of Φ at various pressure levels rather than as contours of pressure on horizontal surfaces. It is usual to express Φ in terms of *geopotential height* Φ/g_0, where g_0 is the standard value of g at the surface (problem 7.11).

7.6   The thermal wind equation

Horizontal pressure gradients arise in the atmosphere owing to density differences which in turn are related to temperature gradients. To relate the geostrophic wind field and its variation with height to the temperature field (7.14) can be applied to two surfaces of constant pressure p_1 and p_2 so that

$$\mathbf{V}_g(p_2) - \mathbf{V}_g(p_1) = \frac{1}{f}\mathbf{k}\wedge\nabla_p(\Phi_2 - \Phi_1) \tag{7.15}$$

Now, from the hydrostatic equation

$$\Phi_2 - \Phi_1 = \frac{R\overline{T}}{M_r}\ln\frac{p_1}{p_2} \tag{7.16}$$

where $\overline{T}$ is the mean temperature between the surfaces so that

$$\mathbf{V}_g(p_2) - \mathbf{V}_g(p_1) = \mathbf{V}_t = \frac{R}{M_r f}\ln\frac{p_1}{p_2}\mathbf{k}\wedge\nabla_p\overline{T} \tag{7.17}$$

The quantity $\mathbf{V}_t$ is called the *thermal wind* over the interval (p_1, p_2); it is related to the $\overline{T}$ field in a similar way to which $\mathbf{V}_g$ is related to the Φ field (fig. 7.5). The quantity $\Phi_2 - \Phi_1$ is known as the *thickness*; thickness charts

Fig. 7.5

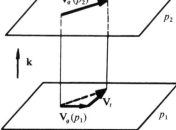

Fig. 7.6

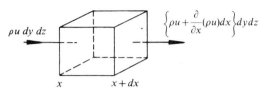

are effectively charts of the mean temperature between different pressure levels.

Various illustrations of the use of thermal wind are given in the problems at the end of the chapter. Fig. 5.1 shows the mean fields of zonal wind and temperature illustrating, for the atmosphere as a whole, how the two are related.

7.7   The equation of continuity

A further equation we shall require in later chapters is the equation of continuity which states that the net flow of mass into unit volume per unit time is equal to the local rate of change of density.

In the elementary volume (fig. 7.6) where the density is ρ and u, v, w the velocity components along Cartesian axes, the mass entering the volume per unit time at x over the face of area $dy\ dz$ is $\rho u\ dy\ dz$, and leaving at $x + dx$ is $(\rho u + (\partial/\partial x)(\rho u)dx)\ dy\ dz$. Adding all the components together, we can write,

$$\frac{\partial(\rho u)}{\partial x} + \frac{\partial(\rho v)}{\partial y} + \frac{\partial(\rho w)}{\partial z} = -\frac{\partial \rho}{\partial t}$$

or $\operatorname{div} \rho \mathbf{V} = -\dfrac{\partial \rho}{\partial t}$ (7.18)

Equation (7.18) is the *equation of continuity*. For an incompressible fluid it reduces to

$\operatorname{div} \mathbf{V} = 0$ (7.19)

A useful form of the continuity equation in which the pressure p is the vertical co-ordinate is found by considering an elemental column of atmosphere (fig. 7.7) confined between the surfaces of constant pressure p and $p - \delta p$. The mass of the column δm is equal to $\rho\ \delta x\ \delta y\ \delta z$ which by the hydrostatic equation gives

$$\delta m = \frac{\delta x\ \delta y\ \delta p}{g}$$

Fig. 7.7

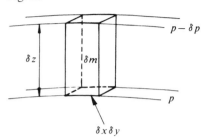

Following the motion, the mass δm is conserved i.e.

$$0 = \frac{1}{\delta m}\frac{d}{dt}(\delta m) = \frac{g}{\delta x\,\delta y\,\delta p}\frac{d}{dt}\left(\frac{\delta x\,\delta y\,\delta p}{g}\right)$$

Carrying out the differentiation and taking the limits we find

$$\frac{\partial u}{\partial x} + \frac{\partial v}{\partial y} + \frac{\partial \omega}{\partial p} = 0 \tag{7.20}$$

where $\omega = dp/dt$.

Equation (7.20) is the continuity equation in isobaric co-ordinates; note that, as with many other equations written in the isobaric system, it does not involve the density.

Problems

7.1 Calculate the value of the centripetal acceleration of a particle at the equator and compare with g. What is the deviation of the plumb-line direction from the true direction of the centre of a uniform spherical earth at latitude 45°?

7.2 Show that

$$\frac{D\mathbf{V}}{Dt} = \mathbf{i}\frac{Du}{Dt} + \mathbf{j}\frac{Dv}{Dt} + \mathbf{k}\frac{Dw}{Dt}$$
$$+ u\frac{D\mathbf{i}}{Dt} + v\frac{D\mathbf{j}}{Dt} + w\frac{D\mathbf{k}}{Dt} \tag{7.21}$$

Show that for the set of axes defined by fig. 7.2

$$\frac{D\mathbf{i}}{Dt} = u\frac{\partial \mathbf{i}}{\partial x} = \frac{u}{a\cos\phi}(\mathbf{j}\sin\phi - \mathbf{k}\cos\phi) \tag{7.22}$$

$$\frac{D\mathbf{j}}{Dt} = u\frac{\partial \mathbf{j}}{\partial x} + v\frac{\partial \mathbf{j}}{\partial y} = -\frac{u\tan\phi}{a}\mathbf{i} - \frac{v}{a}\mathbf{k} \tag{7.23}$$

$$\frac{D\mathbf{k}}{Dt} = u\frac{\partial \mathbf{k}}{\partial x} + v\frac{\partial \mathbf{k}}{\partial y} = \frac{u}{a}\mathbf{i} + \frac{v}{a}\mathbf{j} \tag{7.24}$$

Hence derive (7.6).

7.3 For synoptic scale motions of horizontal dimension $\sim 10^3$ km, of vertical dimension ~ 10 km for which a typical horizontal velocity is $10\,\mathrm{m\,s^{-1}}$, a typical vertical velocity is $1\,\mathrm{cm\,s^{-1}}$, estimate the relative magnitudes of the various terms in (7.6) and (7.7). Compare the magnitude of the terms in the vertical direction with g.

7.4 What is the pressure gradient required at the earth's surface at $45°$ latitude to maintain a geostrophic wind velocity of $30\,\mathrm{m\,s^{-1}}$?

7.5 For motion around a centre of pressure 100 km away, at $30°$ latitude, compute the wind speed for which the Coriolis term will be equal to the acceleration term.

7.6 In the absence of a pressure gradient show that the radius of curvature of the flow is $-V/f$ and is anticyclonic. What is the period of a complete oscillation? Such flow is known as *inertial* flow and has been observed in the oceans as well as in the atmosphere.

7.7 Compute the value of the Rossby number for a typical case (latitude $45°$, $L \simeq 1000$ km, $V \simeq 10\,\mathrm{m\,s^{-1}}$). For the same case, what is the value of the Rossby number on Mars and Venus?

7.8 When $Ro = 0.1$ what is the error in the geostrophic wind approximation?

7.9 When friction is neglected in (7.8) show that when the other terms are retained, in the presence of a pressure gradient $\partial p/\partial r$ for flow with radius of curvature r,

$$V = -\frac{fr}{2} \pm \left[\frac{f^2 r^2}{4} + \frac{r}{\rho}\frac{\partial p}{\partial r}\right]^{1/2} \tag{7.25}$$

This approximation is known as the *gradient wind*. Note that r is measured from the centre of curvature and V is considered positive when cyclonic and negative when anticyclonic.

Show that for anticyclonic curvature

$$\left|\frac{\partial p}{\partial r}\right| < \frac{\rho r f^2}{4} \tag{7.26}$$

and hence that for anticyclones the pressure gradient decreases towards the centre. This is why the pressure gradients are small and the winds light near the centre of an anticyclone.

Draw diagrams showing the balance of forces for the two cyclonic and two anticyclonic solutions of equation (7.25) and comment on their physical realizability.

7.10 The atmospheric surface pressure at radius r_0 from the centre of a tornado rotating with constant angular velocity ω is p_0. Show that the surface pressure at the centre of the tornado is

$p_0 \exp(-\omega^2 r_0^2 M_r/2RT)$

where the temperature T is assumed constant.

7.11 The geopotential height z_g is defined such that

$$z_g = \frac{\Phi}{g_0} = \frac{1}{g_0}\int_0^z g\,dz \tag{7.27}$$

where Φ is the geopotential and where g_0 has a standard value of $9.807\,\mathrm{m\,s^{-2}}$, its mean value at the surface.

Derive an expression for the variation with height of the acceleration due to gravity. Over a place where the value of g at the surface $(z = 0)$ is equal to g_0, what is the difference between z_g and the geometric height z when $z = 100\,\mathrm{km}$?

The value of g decreases by approximately 0.5% between the equator and the poles. Again for $z = 10\,\mathrm{km}$ and $100\,\mathrm{km}$, compute the differences in geopotential height z_g and geometric height z, over the equator and over the poles (cf. appendix 4).

7.12 On the 70 kPa surface at latitude 45° the contours for 3300 m and 3500 m geopotential height are 1000 km apart. What is the magnitude of the geostrophic wind?

7.13 Fig. 7.8(a) is a cross-section of part of a front. At a certain level, the air possesses temperatures T_2 and T_1 respectively $(T_2 > T_1)$ on the two sides of the front which has a slope of α relative to the horizontal. Apply the hydrostatic equation and the geostrophic wind equation to the region AB, together with the condition that the pressure must be continuous across the frontal surface. Show that in equilibrium the components of velocity v_2 and v_1 in the y direction on the two sides of the front satisfy the relation

$$(T_1 - T_2)g \tan \alpha = (v_2 T_1 - v_1 T_2)f \tag{7.28}$$

Fig. 7.8. (a) Cross-section through a front. (b) Warm sector of typical depression showing warm and cold fronts.

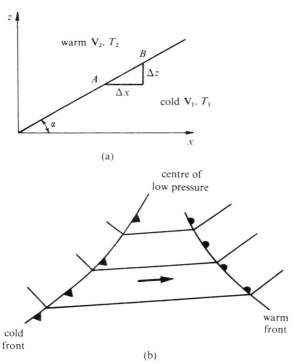

(a)

(b)

If $T_2 - T_1 = 3\,\text{K}$, $v_1 - v_2 = 10\,\text{m s}^{-1}$, find α. The slopes of typical fronts vary from around $\frac{1}{50}$ to $\frac{1}{300}$. Note that if $f = 0$, i.e. the earth were not rotating, the sloping surface could not be in equilibrium.

The cloud features associated with a front are illustrated in fig. 7.9.

7.14 From the sense of the velocity change from (7.28), show that the kink in the isobars is always cyclonic (fig. 7.8(b)).

7.15 The wind at the surface is from the west. At cloud level it is from the south. Do you expect the temperature to rise or fall?

7.16 If the pole is 40 K colder than the equator and the surface wind is zero what wind would you expect at the 20 kPa pressure level? Compare the temperature and wind fields of fig. 5.1 and show so far as you can that they satisfy (7.17).

7.17 At all levels from 100 kPa to 30 kPa for the front shown in fig. 7.8(a) assume a temperature difference $T_2 - T_1$ of 5 K. For 45° latitude

Fig. 7.9. This very high resolution visible image over western Europe was taken by the satellite NOAA-9 at 1324 on 3 October 1985 and received at the University of Dundee Electronics Laboratory. An anticyclone with mainly clear skies was drifting away towards south-eastern Europe, while a deep depression, responsible for the swirl of cloud in the top left hand corner of the picture, was moving slowly from the Atlantic towards northern Britain. The conspicuous cloud band from the British Isles to western France and north-west Spain marks an eastward-moving cold front that separated very warm subtropical air that covered most of Europe from cool polar maritime air over western Britain and the Atlantic.

The bright (and therefore thick) band of cloud associated with the cold front over extreme western Europe results from the ascent of the warm subtropical air over the cooler Atlantic air. It is mostly layered, consisting of water droplets and ice crystals, and produced an area of rain. Embedded within the cloud mass, particularly over northern England, there are some clouds with lumpy appearance, indicating cumuliform development where the air is convectively unstable. The area of cloud ahead of the cold front over Scandinavia is formed by more widespread ascent of the subtropical air.

Immediately behind the front is a largely cloud-free zone where the cold air is subsiding. Further west, in the same cold air mass, there are widespread convective clouds producing showers. Here the polar air is colder than the sea and the air is unstable. Just to the west of Scotland is a 'comma-shaped' cluster of convective cells associated with a region of locally enhanced ascent.

apply the thermal wind equation to the region between two places $x = 0$ and $x = 500\,\text{km}$, the front being at the surface at $x = 0$ and at the 30 kPa level at $x = 500\,\text{km}$. If the surface wind is zero what wind speed will occur at the 30 kPa level? This rather crude calculation shows that the presence of high upper level winds in the vicinity of a front roughly parallel to the frontal surface is consistent with the thermal wind equation. The concentrated region of high winds in the upper troposphere associated with fronts is known as the *jet stream*.

7.18 For a horizontal temperature gradient in the y direction only, derive a differential form of the thermal wind equation (7.17) giving the shear with height of the wind component along the x axis, i.e.

$$\frac{\partial u}{\partial z} = \frac{g}{Tf}\frac{\partial T}{\partial y} \qquad (7.29)$$

7.19 For horizontal flow (i.e. $w = 0$) show that

$$\text{div}_h \mathbf{V} = \nabla_h \cdot \mathbf{V} = \left(\mathbf{i}\frac{\partial}{\partial x} + \mathbf{j}\frac{\partial}{\partial y}\right)\cdot(\mathbf{i}u + \mathbf{j}v)$$

Hence using some of the results of problem 7.2 show that

$$\text{div}_h \mathbf{V} = \frac{\partial u}{\partial x} + \frac{\partial v}{\partial y} - \frac{v}{a}\tan\phi \qquad (7.30)$$

7.20 From (7.30) show that if ρ is considered constant the divergence of the geostrophic wind $\mathbf{V}_g$ is

$$\text{div}_h \mathbf{V}_g = -\frac{v_g}{a}(\tan\phi + \cot\phi) \qquad (7.31)$$

7.21 For the co-ordinate system of fig. 7.2 show that

$$\text{div }\mathbf{V} = \frac{\partial u}{\partial x} + \frac{\partial v}{\partial y} - \frac{v\tan\phi}{a} + \frac{\partial w}{\partial z} + \frac{2w}{a} \qquad (7.32)$$

Show from typical values that the last term $2w/a$ may be neglected in comparison with $\partial w/\partial z$.

7.22 From (7.10) and (7.30) calculate an expression for the divergence of the geostrophic wind. Show by putting in typical values that the terms including $(\partial/\partial x)(1/\rho)$, $(\partial/\partial y)(1/\rho)$ are small compared with the others. Also show, again by inserting typical values, that div $\mathbf{V}_g$ is an order of magnitude smaller than either of the terms $\partial u/\partial x$, $\partial v/\partial y$.

Fig. 7.10. Zonal mean temperature from the surface of Venus to 100 km altitude from Pioneer Venus. Below 55 km altitude the data is from the entry probes and above that altitude from the VORTEX instrument on the orbiter. The reference pressure p_0 for the ordinate is 9216 kPa. (After Schofield & Taylor, 1983)

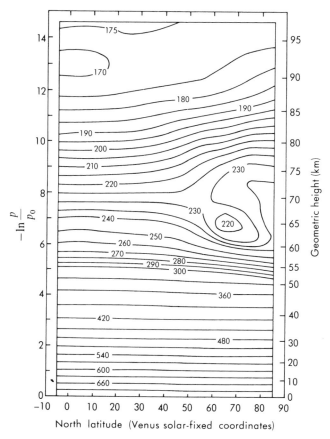

North latitude (Venus solar-fixed coordinates)

7.23 From (7.1) and (7.18) show that an alternate form of the continuity equation is

$$\frac{D\rho}{Dt} + \rho \operatorname{div} \mathbf{V} = 0 \tag{7.33}$$

7.24 Show that when the pressure p is used as the vertical co-ordinate, for a scalar ψ

$$\frac{D\psi}{Dt} = \frac{\partial \psi}{\partial t} + u\frac{\partial \psi}{\partial x} + v\frac{\partial \psi}{\partial y} + \frac{dp}{dt}\cdot\frac{\partial \psi}{\partial p} \tag{7.34}$$

7.25 Starting with (7.18) and using the hydrostatic equation derive the continuity equation in isobaric co-ordinates (7.20) by algebraic manipulation.

7.26 Omitting the frictional force $\mathbf{F}$, multiply (7.4) by ρ and take its curl to obtain the *vorticity equation*

$$\boldsymbol{\nabla} \wedge \left(\rho \frac{D\mathbf{V}}{Dt} \right) = \boldsymbol{\nabla} \wedge (2\rho\mathbf{V} \wedge \boldsymbol{\Omega}) - \mathbf{g} \wedge \boldsymbol{\nabla}\rho \tag{7.35}$$

Hence show that a state of hydrostatic equilibrium (i.e. $\mathbf{V} = 0$ everywhere) is impossible unless there are no variations of density on surfaces of constant geopotential. This result is known as the *Bjerknes–Jeffreys theorem*.

7.27 Venus is a planet having a very small rate of rotation about its axis. Its upper atmosphere possesses a strong zonal motion. Assuming zonal flow u, no meridional flow (i.e. $v = 0$), hydrostatic equilibrium in the vertical and a negligible Coriolis term, show that the equation for cyclostrophic balance is

$$u^2 \tan\phi = -\frac{1}{\rho}\left(\frac{\partial p}{\partial \phi} \right)_z \tag{7.36}$$

p and ρ being the pressure and density at latitude ϕ and altitude z. Note that for cyclostrophic balance to hold the poleward pressure gradient must be less than zero.

Adopting pressure as the vertical co-ordinate, show that

$$u^2(p,\phi) = -\frac{R}{M \tan\phi} \int_{p_0}^{p} \left(\frac{\partial T}{\partial \phi} \right)_p d(\ln p) + u^2(p_0,\phi)$$

From the temperature information of fig. 7.10 assuming $u(p, \phi) = 0$ at an altitude of 50 km, calculate the field of $u(p, \phi)$ for altitudes between 50 and 70 km and for ϕ between $30°$ and $65°$. What is the maximum velocity of the jet?

8

Atmospheric waves

8.1 Introduction

In chapter 7 we dealt with various approximations to the equations of motion, particularly those applicable to the fairly large scale. No attempt was made to consider how different forms of motion might arise. Before approaching a discussion of why the circulation of the atmosphere is as we find it, it is instructive to look at simple but important types of wave motion which are present in the atmosphere. To isolate simple wave forms we shall again make severe approximations.

For each case the appropriate equations are (1) the momentum equations (§7.2), (2) the continuity equation (§7.7) and (3) the first law of thermodynamics. Our method of solution will be the perturbation method.

8.2 Sound waves

To illustrate the method we first consider sound waves. For a simple treatment of these it will suffice to assume motion along the x direction only and that the y and z components and gradients in these directions are zero. The Coriolis term and friction are also omitted and the motion is assumed adiabatic. The momentum equation (7.8) becomes

$$\frac{Du}{Dt} + \frac{1}{\rho}\frac{\partial p}{\partial x} = 0 \tag{8.1}$$

the continuity equation (7.18) becomes

$$\frac{D\rho}{Dt} + \rho\frac{\partial u}{\partial x} = 0 \tag{8.2}$$

and the first law of thermodynamics for adiabatic motion is

$$p\rho^{-\gamma} = \text{constant} \tag{8.3}$$

where γ is the ratio of specific heats of dry air. Combining (8.2) and (8.3) we obtain

$$\frac{1}{\gamma} \frac{D \ln p}{Dt} + \frac{\partial u}{\partial x} = 0 \qquad (8.4)$$

The perturbation method consists of allowing the variables to be written as the sum of a mean component (represented by a bar) and a variable component (represented by a prime)

i.e.
$$u = \bar{u} + u'$$
$$p = \bar{p} + p' \qquad (8.5)$$
$$\rho = \bar{\rho} + \rho'$$

Here $p' \ll \bar{p}$, $\rho' \ll \bar{\rho}$ but u' is not necessarily $< \bar{u}$ as the solutions are still valid if $\bar{u} = 0$. Substituting in (8.1) and (8.4), neglecting products of primed quantities, assuming that differentials of mean quantities are zero, and also using (7.1) we find

$$\left(\frac{\partial}{\partial t} + \bar{u} \frac{\partial}{\partial x} \right) u' + \frac{1}{\bar{\rho}} \frac{\partial p'}{\partial x} = 0 \qquad (8.6)$$

$$\left(\frac{\partial}{\partial t} + \bar{u} \frac{\partial}{\partial x} \right) p' + \gamma \bar{p} \frac{\partial u'}{\partial x^2} = 0 \qquad (8.7)$$

Eliminating u' between these equations,

$$\left(\frac{\partial}{\partial t} + \bar{u} \frac{\partial}{\partial x} \right)^2 p' - \frac{\gamma \bar{p}}{\bar{\rho}} \frac{\partial^2 p'}{\partial x^2} = 0 \qquad (8.8)$$

which is a wave equation having solutions

$$p' = \text{Re}\{A \exp ik(x - ct)\} \qquad (8.9)$$

with a constant A and

$$c = \bar{u} \pm (\gamma \bar{p}/\bar{\rho})^{1/2} \qquad (8.10)$$

The speed of sound wave relative to the flow $\bar{u}$ is therefore $(\gamma \bar{p}/\bar{\rho})^{1/2}$.

8.3   Gravity waves

Since sound waves are longitudinal, only one dimension need be considered to obtain the sound wave solution. Other waves exist where the oscillation is transverse to the direction of propagation; some of these waves are known as *gravity waves*. Their existence is illustrated by the discussion of §1.4 where we considered the stability of a parcel displaced vertically from its equilibrium level. In problem 1.10 for a stable stratification the frequency of oscillation (the Brunt–Vaisala frequency) of such a parcel was found on the assumption that the environment is unaffected by the motion – a simple calculation which applies in the limiting case of large vertical scale and small horizontal scale.

To obtain a reasonably simple solution to the basic equations for gravity waves we shall (1) work in two dimensions and ignore motion or gradients along the y direction, (2) assume that the horizontal scale of the wave is sufficiently small that the Coriolis term may be neglected by comparison with the other terms, (3) ignore the friction term, (4) assume that the unperturbed atmosphere is at rest. The components of the momentum equation (7.4) are

$$\frac{Du}{Dt} + \frac{1}{\rho}\frac{\partial p}{\partial x} = 0 \tag{8.11}$$

$$\frac{Dw}{Dt} + \frac{1}{\rho}\frac{\partial p}{\partial z} + g = 0 \tag{8.12}$$

The continuity equation (7.33) becomes

$$\frac{1}{\rho}\frac{D\rho}{Dt} + \frac{\partial u}{\partial x} + \frac{\partial w}{dz} = 0 \tag{8.13}$$

and the first law of thermodynamics for adiabatic motion means that the potential temperature θ remains constant, i.e.

$$\frac{D\ln\theta}{Dt} = 0 \tag{8.14}$$

From an equation similar to (3.3), θ can be expressed in terms of ρ and p such that

$$\frac{D\ln\theta}{Dt} = \frac{1}{\gamma}\frac{D\ln p}{Dt} - \frac{D\ln\rho}{Dt} \tag{8.15}$$

For the unperturbed atmosphere neither θ, p nor ρ vary in the horizontal.

Let perturbations be introduced in (8.11), (8.12), (8.13) and (8.15), such that $u = u'$, $w = w'$, $\rho = \bar{\rho} + \rho'$, and $p = \bar{p} + p'$. Then using (7.1), the hydrostatic equation $\bar{p}^{-1}(\partial\bar{p}/\partial z) = -H^{-1}$, the equation of state $\bar{p}/\bar{\rho} = gH$, where H is the scale height, also neglecting products of primed quantities, the four equations (8.11), (8.12), (8.13) and (8.15) become for the unknowns u', w', $\rho'/\bar{\rho}$ and $p'/\bar{p}$ (problem 8.2):

$$\frac{\partial u'}{\partial t} + gH\frac{\partial}{\partial x}\left(\frac{p'}{\bar{p}}\right) = 0 \tag{8.16}$$

$$\frac{\partial w'}{\partial t} + g\left(\frac{\rho'}{\bar{\rho}}\right) + gH\frac{\partial}{\partial z}\left(\frac{p'}{\bar{p}}\right) - g\left(\frac{p'}{\bar{p}}\right) = 0 \tag{8.17}$$

$$\frac{\partial u'}{\partial x} + \frac{\partial w'}{\partial z} - \frac{w'}{H} + \frac{\partial}{\partial t}\left(\frac{\rho'}{\bar{\rho}}\right) = 0 \tag{8.18}$$

$$-\frac{N_B^2}{g}w' + \frac{\partial}{\partial t}\left(\frac{\rho'}{\bar{\rho}}\right) - \frac{1}{\gamma}\frac{\partial}{\partial t}\left(\frac{p'}{\bar{p}}\right) = 0 \tag{8.19}$$

where, according to the nomenclature of problem 1.10, the vertical stability parameter $\frac{N_B^2}{g}(= B)$ has been written for $\partial \ln \theta / \partial z$.

In solving these equations we shall assume first an isothermal atmosphere in which H is a constant, in which case $N_B^2/g = (\gamma - 1)/\gamma H$ and is, of course, also constant. We look for wave solutions such that each of the four unknowns vary as

$$\exp(\alpha z)\exp i(\omega t + kx + mz)$$

in which the first exponential has been introduced to allow for variation of amplitude with altitude. Solutions of this kind will exist if after substituting them into (8.16) to (8.19) the determinant of the coefficients is equal to zero. On equating imaginary parts we find that unless $m = 0$, $\alpha = 1/2H$. Solutions with the first condition have no phase variation in the vertical and are known as *external waves*; the most important of these are *surface waves* whose energy is concentrated at a boundary or discontinuity in the atmosphere similar to surface waves on the ocean. When $m \neq 0$, we have *internal waves*. Substituting $\alpha = 1/2H$ into the equation and equating the real part of the determinant of the coefficients to zero the following dispersion relation results (problem 8.3):

$$m^2 = k^2\left(\frac{N_B^2}{\omega^2} - 1\right) + \frac{(\omega^2 - \omega_a^2)}{c^2} \tag{8.20}$$

where c is the velocity of sound,

$$N_B^2 = \frac{(\gamma - 1)g}{\gamma H} \tag{8.21}$$

is the square of the Brunt–Vaisala frequency for the isothermal atmosphere (problem 8.4) and

$$\omega_a = \frac{1}{2}\left(\frac{\gamma g}{H}\right)^{1/2} = \frac{c}{2H} \tag{8.22}$$

Fig. 8.1. Dispersion curves for gravity waves. The full lines correspond to $m = 0$, the dashed line is for a wave having the velocity of sound c.

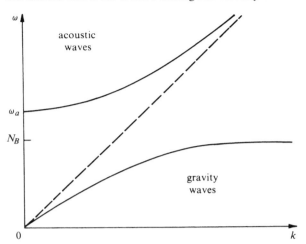

is known as the *acoustic cut-off frequency*.

The dispersion relation (8.20) is illustrated in fig. 8.1 where curves of $m = 0$ are plotted on a ω–k diagram. Two regions where m^2 is positive are apparent, the higher frequency region ($\omega > \omega_a$ when $k = 0$) describes acoustic waves and the lower frequency region, gravity waves. Since $N_B < \omega_a$ (problem 8.6) these two regions are well separated. Note also from (8.20) that if $m \ll k$, i.e. for deep waves of short horizontal wavelength compared with vertical wavelength, $\omega \simeq N_B$ as would be expected from our simple derivation of N_B in problem 1.10.

If the assumption that the atmosphere is isothermal is removed, (8.20) may still be employed as a reasonable approximation with N_B as the Brunt–Vaisala frequency for the atmosphere in question, provided $N_B < \omega_a$ as is normally the case throughout the atmosphere (problem 8.6).

If the atmosphere is in a state of uniform zonal flow $\bar{u}$, (8.20) still holds provided ω is replaced by $\omega + \bar{u}k$, the frequency which would be seen by an observer moving with the basic flow and known as the *intrinsic frequency*.

Equation (8.20) applies to waves having a wide range of wavelength, frequency and velocity. For gravity waves of horizontal wavelength of the order of a few kilometres, the first term is much larger than the second. In this case, allowing also for the presence of a uniform zonal wind $\bar{u}$ the dispersion relation (8.20) may be written

Fig. 8.2. The structure of lee waves according to Gerbier & Berenger (1961).

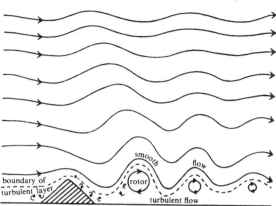

$$\frac{(\omega - \bar{u}k)^2}{N_B^2} = \frac{k^2}{m^2 + k^2} \qquad (8.23)$$

The most easily observed example of such gravity waves is that of waves in the lee of mountains. Air which is forced to flow over a mountain under stable conditions is set into a gravity-wave oscillation as it moves downstream from the mountain. If the amplitude is large enough and the conditions of temperature and humidity suitable, clouds will form in the regions where the air has been lifted (figs. 3.4 and 3.5). Such waves will, of course, be stationary with respect to the mountain and the horizontal component of phase velocity relative to the surface will be zero, in which case for $k \ll m$ from (8.23) we have

$$m = N_B/\bar{u} \qquad (8.24)$$

Since the energy source is near the surface, energy is propagated upwards (see Problem 11.5). Their group velocity is, therefore, upwards and their phase velocity downwards (problem 8.10). Consequently the lines of constant phase in the stationary waves tilt with height backwards relative to the mean flow (fig. 8.2), although in practice, because the mean wind varies with height, the situation is rarely so simple as this.

Gravity waves represent only a minor component of the motion in the lower atmosphere. Above 75 km, however, atmospheric motion is probably dominated by them (fig. 8.3). Gravity waves include tidal motions, tides being a special case of gravity waves having a particular horizontal scale and a particular period. These motions show up for instance in the variations of ionized layers as well as in winds determined from the movement of trails left by rockets or by meteors. Although mainly generated in

Fig. 8.3. Profiles of the W–E component of the wind as measured from 36 rockets which carried chemical trail experiments all near 30°N latitude between 15 November and 14 December or near 30°S latitude between 15 May and 14 June. (From COSPAR 1972)

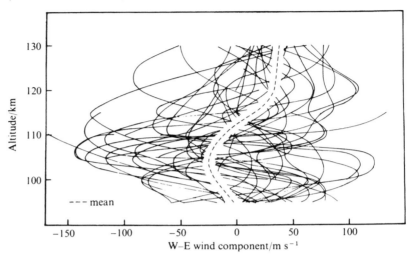

the lower atmosphere gravity waves are propagated upwards, where as the density decreases their amplitude increases (problem 8.8). At levels above 75 km or so they are dissipated by viscous damping. Because of this they contribute significantly to the energy budget of the upper mesosphere and lower thermosphere. They also play a critical role in determining the momentum budget in the region of the mesopause (§10.9).

8.4   Rossby waves

The very large scale wave features which are observed in the flow on the planetary scale are known as *planetary waves* (fig. 8.4). In their simplest form they occur because of the variation of the Coriolis parameter with latitude and are known as *Rossby waves*.

The simplest Rossby wave solution is obtained for an atmosphere of constant density under the assumption of uniform zonal flow $\bar{u}$ and no vertical motion. The appropriate momentum equations (7.8) and the continuity equation (7.19) are:

$$\frac{Du}{Dt} + \frac{1}{\rho}\frac{\partial p}{\partial x} - fv = 0 \tag{8.25}$$

$$\frac{Dv}{Dt} + \frac{1}{\rho}\frac{\partial p}{\partial y} + fu = 0 \tag{8.26}$$

Fig. 8.4. Geopotential height of the 50 kPa pressure surface in decametres for a typical northern hemisphere winter situation. The large scale waves are known as *planetary waves*.

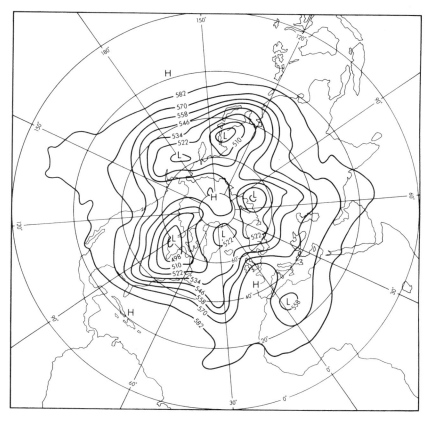

$$\frac{\partial u}{\partial x} + \frac{\partial v}{\partial y} = 0 \qquad (8.27)$$

Operating on (8.25) with $\partial/\partial y$ and (8.26) with $\partial/\partial x$ and subtracting

$$\frac{D}{Dt}\left(\frac{\partial v}{\partial x} - \frac{\partial u}{\partial y}\right) + f\left(\frac{\partial v}{\partial y} + \frac{\partial u}{\partial x}\right) + v\frac{\partial f}{\partial y} = 0 \qquad (8.28)$$

The quantity $(\partial v/\partial x) - (\partial u/\partial y)$ is one component of curl **V** which is known as the *vorticity* ζ; it may be considered as a vector equal to twice the local angular velocity of fluid elements. The second term in (8.28) is zero by (8.27). Because the Coriolis parameter f only varies with latitude it is possible to write (8.28) in the form

$$\frac{D(\zeta + f)}{Dt} = 0 \qquad (8.29)$$

The quantity $\zeta + f$ is known as the *absolute vorticity*; it is the vorticity due to the rotation of the fluid itself combined with that due to the earth's rotation. Equation (8.29) shows that under the conditions we have imposed of non-divergent, frictionless flow, absolute vorticity is conserved.

To solve (8.29) we assume a linear relation between f and y, i.e. we let $f = f_0 + \beta y$ where β is a constant (this is known as the *β-plane approximation*). We assume a uniform zonal flow $\bar{u}$ in the unperturbed situation and introduce perturbations, i.e. $u = \bar{u} + u'$, $v = v'$ so that (8.29) becomes

$$\left(\frac{\partial}{\partial t} + \bar{u} \frac{\partial}{\partial x} \right) \left(\frac{\partial v'}{\partial x} - \frac{\partial u'}{\partial y} \right) + \beta v' = 0 \qquad (8.30)$$

Because the flow is non-divergent in the horizontal a *stream function* ψ can be introduced such that by using it (8.27) is automatically satisfied,

i.e. $$u' = -\frac{\partial \psi}{\partial y}, \quad v' = \frac{\partial \psi}{\partial x} \qquad (8.31)$$

Substituting in (8.30),

$$\left(\frac{\partial}{\partial t} + \bar{u} \frac{\partial}{\partial x} \right) \nabla^2 \psi + \beta \frac{\partial \psi}{\partial x} = 0 \qquad (8.32)$$

For wave solutions $\psi = \mathrm{Re}\{\psi_0 \exp i(\omega t + kx + ly)\}$ to be possible the dispersion relation

$$c = -\frac{\omega}{k} = \bar{u} - \frac{\beta}{k^2 + l^2} \qquad (8.33)$$

must be satisfied. The velocity relative to the zonal flow is $c - \bar{u}$ where c is the phase velocity in the x direction. Rossby waves, therefore, drift to the west relative to the basic flow, at typical speeds of a few metres per second (problem 8.14). Note that the phase speed of the waves increases with wavelength.

8.5   The vorticity equation

Under the assumptions of two dimensional flow in an atmosphere of uniform density (8.29) is a statement of the conservation of absolute vorticity. Before dealing with the propagation of Rossby-type waves in three dimensions we need to find an equation to describe vorticity changes under less stringent assumptions. For motions of synoptic scale this can be done by starting with the approximate horizontal momentum equations

(8.25) and (8.26). Operating on (8.25) with $\partial/\partial y$ and (8.26) with $\partial/\partial x$ and subtracting, also noting that $Df/Dt = v\partial f/\partial y$ we have for the rate of change of absolute vorticity

$$\frac{D}{Dt}(\zeta + f) = -(\zeta + f)\left(\frac{\partial u}{\partial x} + \frac{\partial v}{\partial y}\right)$$
$$-\left(\frac{\partial w}{\partial x}\frac{\partial v}{\partial z} - \frac{\partial w}{\partial y}\frac{\partial u}{\partial z}\right) + \frac{1}{\rho^2}\left(\frac{\partial \rho}{\partial x}\frac{\partial p}{\partial y} - \frac{\partial \rho}{\partial y}\frac{\partial p}{\partial x}\right) \qquad (8.34)$$

Equation (8.34) is known as the *vorticity equation* (cf. (7.35)). The first term on the right hand side is the most important; it arises because of the horizontal divergence. If there is positive horizontal divergence, air is flowing out of the region in question, and the vorticity decreases. This is the same effect as occurs with a rotating body whose angular velocity decreases because of angular momentum conservation if its moment of inertia increases.

Scale analysis of (8.34) shows that for synoptic scale motions the last two terms are considerably smaller than the others (problem 8.15) and that to a first approximation

$$\frac{D_h}{Dt}(\zeta + f) = -(\zeta + f)\left(\frac{\partial u}{\partial x} + \frac{\partial v}{\partial y}\right) \qquad (8.35)$$

where D_h/Dt denotes

$$\frac{\partial}{\partial t} + u\frac{\partial}{\partial x} + v\frac{\partial}{\partial y}$$

It is instructive to apply (8.35) to an atmosphere of constant density and temperature for which the continuity equation is (7.19) so that (8.35) becomes

$$\frac{D_h}{Dt}(\zeta + f) = (\zeta + f)\frac{\partial w}{\partial z} \qquad (8.36)$$

Because of the constant temperature the geostrophic wind is independent of the height z. Further, because to a first approximation the vorticity is equal to the vorticity of the geostrophic wind the vorticity will not vary with height (problem 8.16). Therefore, integrating (8.36) between levels z_1 and z_2 where $z_2 - z_1 = h$ we have

$$\frac{1}{(\zeta + f)}\frac{D_h}{Dt}(\zeta + f) = \frac{w(z_2) - w(z_1)}{h} \qquad (8.37)$$

Now considering the fluid which at one time is confined between the levels distance h apart, we have

$$\frac{Dh}{Dt} = w(z_2) - w(z_1) \tag{8.38}$$

so that (8.37) may now be written

$$\frac{D_h}{Dt}\left(\frac{\zeta + f}{h}\right) = 0 \tag{8.39}$$

Equation (8.39) is a simplified statement of the conservation of *potential vorticity*. It has important consequences for atmospheric flow. Consider, for instance, adiabatic flow over a mountain barrier. As a column of air flows over the barrier its vertical extent decreases so that ζ must also decrease. A westward moving airstream will, therfore, move equatorwards as it passes over the barrier (problem 8.17).

8.6   Three-dimensional Rossby-type waves

In §8.4 we assumed two dimensional structure only for the Rossby wave solutions. To find how such waves propagate vertically we need to look for the constraints imposed on three dimensional solutions. To simplify the treatment we still wish to work with an atmosphere at nearly constant density and so we employ the *Boussinesq approximation* which allows us to neglect changes of density except where they are coupled with gravity to produce buoyancy forces. With this approximation the equation of continuity is as for an incompressible fluid, i.e. (7.19) applies, so that the appropriate vorticity equation is (8.36). With a uniform unperturbed zonal flow $\bar{u}$, the perturbation form of (8.36) is

$$\left(\frac{\partial}{\partial t} + \bar{u}\frac{\partial}{\partial x}\right)\zeta' + v'\frac{\partial f}{\partial y} - f\frac{\partial w'}{\partial z} = 0 \tag{8.40}$$

where ζ has been neglected compared with f in the right hand side of (8.36) (problem 8.16).

We also require the perturbation form of the hydrostatic equation:

$$\frac{1}{\bar{\rho}}\frac{\partial p'}{\partial z} + \frac{\rho'}{\bar{\rho}}g = 0 \tag{8.41}$$

and the thermodynamic equation which under the Boussinesq approximation becomes

$$\left(\frac{\partial}{\partial t} + \bar{u}\frac{\partial}{\partial x}\right)\frac{\rho'}{\bar{\rho}} - \frac{N_B^2}{g}w' = 0 \tag{8.42}$$

Comparing (8.42) with (8.19), note that since under the approximation of a nearly incompressible fluid, $\gamma \to \infty$, the last term in (8.19) may be neglected.

Eliminating ρ' and w' from (8.40), (8.41) and (8.42) and putting $\beta = \partial f / \partial y$ (the β-plane approximation) we have

$$\left(\frac{\partial}{\partial t} + \bar{u} \frac{\partial}{\partial x}\right)\left(\zeta' + \frac{f_0}{N_B^2 \bar{\rho}} \frac{\partial^2 p'}{\partial z^2}\right) + \beta v' = 0 \tag{8.43}$$

Now most of the vorticity perturbation arises from the vorticity perturbation of the geostrophic wind (problem 8.16) whose components u_g' and v_g' are

$$u_g' = -\frac{1}{f_0 \bar{\rho}} \frac{\partial p'}{\partial y}, \quad v_g' = \frac{1}{f_0 \bar{\rho}} \frac{\partial p'}{\partial x} \tag{8.44}$$

For the geostrophic wind, as in (8.31), we can define a stream function ψ which, comparing with (8.44), will be equal to $p'/f_0\bar{\rho}$. Introducing this into (8.43) and also approximating v' by its geostrophic value in the last term, (8.43) becomes

$$\left(\frac{\partial}{\partial t} + \bar{u} \frac{\partial}{\partial x}\right)\left(\nabla^2 \psi + \frac{f_0^2}{N_B^2} \frac{\partial^2 \psi}{\partial z^2}\right) + \beta \frac{\partial \psi}{\partial x} = 0 \tag{8.45}$$

These last substitutions involve the *quasi-geostrophic approximation* in which the wind is replaced by its geostrophic value except in the divergence term. For wave solutions

$$\psi = \text{Re}\{\psi_0 \exp i(\omega t + kx + ly + mz)\}$$

to be possible for (8.45), the dispersion relation

$$c = -\frac{\omega}{k} = \bar{u} - \frac{\beta}{\left(k^2 + l^2 + m^2 f_0^2 / N_B^2\right)} \tag{8.46}$$

must be satisfied. Since $f_0^2/N_B^2 \ll 1$ (problem 8.18), vertical wavelengths are of the order of 1% of horizontal wavelengths.

We might expect some of these large scale waves to be forced by features at the surface, by mountains or large land masses, in which case they would be stationary with respect to the surface. For c to be zero from (8.46)

$$m^2 = \frac{N_B^2}{f_0^2}\left\{\frac{\beta}{\bar{u}} - \left(k^2 - l^2\right)\right\} \tag{8.47}$$

Fig. 8.5. Temperature cross-section (in K) of the stratosphere around the latitude circle 65°N as derived from the Selective Chopper Radiometer on the satellite Nimbus 4 for 4 January 1971 at the peak of a *stratospheric warming*. Note the large amplitude wavenumber one (the wavenumber refers to the number of complete cycles of a variation which occur around a latitude circle) and the westward tilt of the wave with height. (After Houghton, 1972)

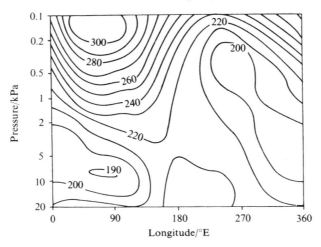

For vertical propagation to occur m^2 must be positive, which sets the condition

$$0 < \bar{u} < \beta/(k^2 + l^2) \tag{8.48}$$

Under conditions of easterly flow, therefore, or very large westerly flow no vertical propagation will occur. This result, first obtained by Charney & Drazin (1961), is confirmed by the lack of planetary wave activity in the stratosphere during the summer months when the mean stratospheric flow is easterly. Further the waves of largest horizontal wavelength propagate most readily. Figs. 8.5 and 8.6 illustrate some such waves observed in the stratosphere; where their amplitude is sufficiently large they are known as *stratospheric warmings*.

Problems

8.1 Write down the horizontal equations of motion for an atmosphere in which pressure oscillations are negligible and in which there is no basic flow, i.e.

Fig. 8.6. Illustrating a quasi-stationary wave in southern hemisphere stratosphere of wavenumber 2 as observed by Selective Chopper Radiometer on the Nimbus 4 satellite on 25 September 1971 (after Harwood, 1975). The quantity plotted is temperature in K of a layer ~10km thick centred at (a) ~45km altitude and (b) 15km altitude. Note that there is about half a wavelength of the wave in the vertical between 45 and 15km.

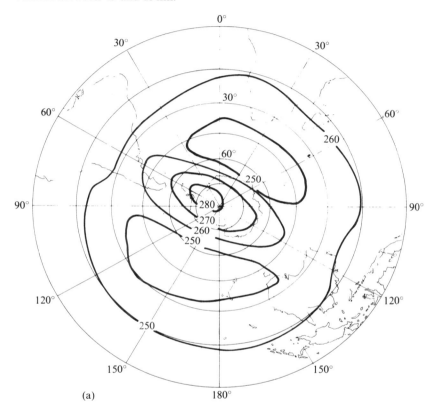

(a)

$$\frac{\partial u}{\partial t} - fv = 0$$

$$\frac{\partial v}{\partial t} + fu = 0 \qquad\qquad (8.49)$$

Substituting $V = u + iv (i = \sqrt{-1})$ solve the resulting equation on the assumption that f is constant and show that *inertial oscillations* occur with a period $2\pi/f$ (cf. problem 7.6).

8.2 Derive (8.16) to (8.19) from (8.11) to (8.15).

8.3 Derive (8.20).

8.4 Show that for an isothermal atmosphere

Fig. 8.6. (*continued*)

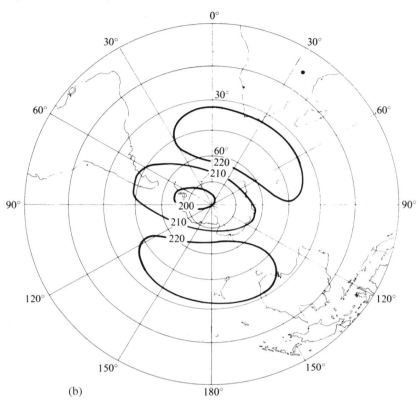

(b)

$$N_B^2/g = (\gamma - 1)/\gamma H \qquad (8.50)$$

(See problem 1.10.)

8.5 With $\bar{u} = 0$, from (8.23), find the horizontal phase velocity of gravity waves for various values of horizontal and vertical wavelength.

8.6 Show that ω_a defined by (8.22) is always greater than N_B for an isothermal atmosphere defined by (8.21). Calculate the value of the lapse rate of temperature with altitude for which the Brunt–Vaisala frequency is equal to ω_a.

8.7 Consider a pattern of motion in which fluid elements move at an angle θ to the vertical. Show that for an atmosphere where the Brunt–Vaisala frequency is N_B the elements will oscillate with a frequency $\omega = N_B \cos \theta$.

Now (following Lindzen, 1971) imagine waves propagating upwards at angle θ to the vertical excited by a surface with uniform corrugations distance $2\pi/k$ apart moving horizontally at the base of the atmosphere. Show that the ratio of horizontal wavelength $(2\pi/k)$ to vertical wavelength $(2\pi/m)$ is $\tan\theta$. Hence deduce the relation (cf. (8.20))

$$m^2 = \frac{1-(\omega/N_B)^2}{(\omega/N_B)^2}k^2$$

8.8 Show that in a wave which is propagating upwards with constant kinetic energy the velocity in the wave will vary with altitude approximately as $\exp(z/2H)$ where H is the scale height. If a gravity wave at 100 km has an amplitude of $100\,\mathrm{m\,s^{-1}}$, what would be its amplitude at the surface where it originates?

8.9 In (8.24) with $\bar{u}=10\,\mathrm{m\,s^{-1}}$, calculate the vertical wavelength for (1) an isothermal atmosphere, (2) an atmosphere with a lapse rate of $5\,\mathrm{K\,km^{-1}}$.

8.10 For gravity waves in an atmosphere with no mean flow ($\bar{u}=0$) and having a dispersion relation given by (8.23) compute the horizontal and vertical components of the group velocity. Show that for $k\ll m$, the vertical component of group velocity is equal and opposite to the vertical component of phase velocity.

8.11 Write down the horizontal momentum equations for perturbations in an atmosphere with no mean flow ($\bar{u}=0$), i.e.

$$\frac{\partial u'}{\partial t}+\frac{1}{\bar{\rho}}\frac{\partial p'}{\partial x}-fv'=0$$

$$\frac{\partial v'}{\partial t}+\frac{1}{\bar{\rho}}\frac{\partial p'}{\partial y}+fu'=0 \tag{8.51}$$

Add (8.17) to (8.19) in their simplified form, assume no variation with y and plane wave solutions as in §8.3. Show that the equations are satisfied and that the appropriate dispersion relation is now

$$\omega^2-f^2=\frac{k^2}{m^2}(N_B^2-\omega^2) \tag{8.52}$$

showing that ω is always $>f$.

This result shows the constraint on ω for gravity waves which are of sufficiently large horizontal extent for the Coriolis term to be important. It is particularly important for consideration of tides. Because f varies from zero at the equator to 2π (12 hours)$^{-1}$ at the poles, internal gravity waves with periods shorter than 12 hours can propagate anywhere but the diurnal tide will be largely confined to latitudes less than 30°.

8.12 From (8.33) calculate the phase velocity of Rossby waves relative to the basic flow for waves moving round latitude 30° of 3000 km latitudinal width and 10000 km wavelength.

8.13 Calculate the zonal wind under which a Rossby wave pattern at 60°N of wavenumber 3 (i.e. 3 maxima around a circle of longitude) and latitudinal width 3000 km would be stationary with respect to the earth's surface.

8.14 From (8.33) find the x component of the group velocity of Rossby waves and when $k > l$ show that relative to the basic flow the group velocity is opposite to the phase velocity.

8.15 For synoptic scale motions in mid latitudes a typical value of horizontal dimension is 1000 km, of horizontal velocity 10 m s^{-1}, of vertical velocity 1 cm s^{-1}. What are corresponding typical values of fractional density variations $\delta\rho/\rho$ and pressure variations $\delta p/p$. Show that the last two terms on the right hand side of (8.34) are about an order of magnitude smaller than the first term on the right hand side.

8.16 Show by scale analysis as in problem 8.15 that in (8.35) the vorticity ζ may be approximated by the vorticity of the geostrophic wind.
    Show also that under the same conditions ζ is considerably smaller than f.

8.17 Consider the implications of the theorem of conservation of potential vorticity applied to the easterly flow over a mountain barrier. A difference arises compared with westerly flow because of the variation of f with latitude. Show that for easterly flow the air must begin to move southward before it reaches the mountain barrier.

8.18 Compute values of f^2/N_B^2 (cf. (8.46)) for several typical atmospheric stabilities (e.g. for an isothermal atmosphere and for an atmosphere having a lapse rate of 5 K km^{-1}).

8.19 Compute the vertical phase velocity and group velocity for waves described by (8.46). Show that for waves in which energy is propagated upwards the phase velocity is downwards and hence that such waves slope towards the west with increasing altitude.

8.20 Write down the horizontal momentum equations (8.51) for perturbations u', v', p' in a form applicable to an equatorial β-plane, i.e. with $f = \beta y$. Assume a solution with $v' = 0$, and

$$u' = \text{Re}\{u_0' \exp i(\omega t - kx)\}$$

Show that in this case the amplitude u_0' varies with latitude as

$$\exp\left(\frac{-\beta y^2 k}{2\omega}\right)$$

Note that for a satisfactory solution k/ω must be positive, i.e. the wave must be *eastward moving*. Plot the pressure field associated with the variation in zonal velocity u'. These are *Kelvin waves*; they have been observed in the equatorial stratosphere.

8.21 For gravity waves under the assumptions of §8.3 with the scale height H large compared with the vertical or horizontal wavelengths of the wave, show that the ratio of the amplitudes of the vertical to the horizontal disturbances in the wave is given by

$$\frac{|w'|}{|u'|} = \frac{-k}{m} \tag{8.53}$$

8.22 From equation (7.1) applied to the velocity **V**, show, by considering individual components or otherwise, that

$$\frac{D\mathbf{V}}{Dt} = \frac{\partial \mathbf{V}}{\partial t} + \zeta \wedge \mathbf{V} + \nabla\left(\frac{1}{2}\mathbf{V}^2\right) \tag{8.54}$$

where $\zeta = \text{curl } \mathbf{V}$ is the vorticity.
From (8.54) and remembering that

$$\nabla \cdot \zeta = 0 \tag{8.55}$$

show that

$$\nabla \wedge \frac{D\mathbf{V}}{Dt} = \frac{D\zeta}{Dt} - (\zeta \cdot \nabla)\mathbf{V} + \zeta(\nabla \cdot \mathbf{V}) \tag{8.56}$$

Combine (8.56) with the equation of continuity (7.33) and show that

$$\frac{1}{\rho}\nabla \wedge \frac{D\mathbf{V}}{Dt} = \frac{D}{Dt}\left(\frac{\zeta}{\rho}\right) - \left(\frac{\zeta}{\rho} \cdot \nabla\right)\mathbf{V} \tag{8.57}$$

Now take the equation of motion (7.2), neglecting the friction **F**, remembering that for any scalar ϕ, $\nabla \wedge \nabla\phi = 0$ and substituting in (8.57) show that

$$\frac{D}{Dt}\left(\frac{\zeta}{\rho}\right) = \left(\frac{\zeta}{\rho} \cdot \nabla\right)\mathbf{V} + \left(\frac{\nabla\rho \wedge \nabla p}{\rho^3}\right) \tag{8.58}$$

Fig. 8.7. Illustrating *baroclinicity*. A density gradient ∇_ρ exists between the heavier fluid on the left and the lighter fluid on the right. A circulation results shown by the arrows. The baroclinicity vector, which according to equation (8.25) is proportional to the rate of change of vorticity, is directed out of the paper. (After Gill, 1982)

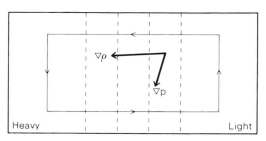

A *baroclinic* situation is one in which horizontal temperature gradients exist so that pressure is not constant on surfaces of constant density or vice versa. The last term in equation (8.58) is called the *baroclinicity vector* since it describes to what extent surfaces of constant density and surfaces of constant pressure are inclined to each other (fig. 8.7).

Take now some property which is conserved by the fluid, for instance, assuming negligible friction, the potential temperature θ, given therefore that

$$\frac{D\theta}{Dt} = 0 \tag{8.59}$$

prove the relation

$$\zeta \cdot \frac{D(\nabla\theta)}{Dt} = -\nabla\theta \cdot (\zeta \cdot \nabla)\mathbf{V} \tag{8.60}$$

Now take the dot product of $\nabla\theta$ with (8.58) and show that

$$\frac{D}{Dt}\left(\frac{\zeta}{\rho} \cdot \nabla\theta\right) = \nabla\theta \cdot \left(\frac{\nabla\rho \wedge \nabla p}{\rho^3}\right) \tag{8.61}$$

Show further that if θ is a function of p and ρ only as is the case, for instance, for a uniform composition inviscid atmosphere, then the right hand side of equation (8.61) is zero.

The quantity

$$\frac{\zeta}{\rho} \cdot \nabla\theta$$

Fig. 8.8. Isopleths of Ertel potential vorticity on an isentropic surface (potential temperature 850 K) near 30 km altitude for 7 December 1981 computed from temperature data from the Stratospheric Sounding Unit on the NOAA 7 satellite. Units are $Km^2 kg^{-1} s^{-1} \times 10^{-4}$. The shaded region has values between 4 and 6 units. Arrows show the geostrophic flow. The structure of the planetary wave suggests that it is 'breaking'. (After Clough et al., 1985)

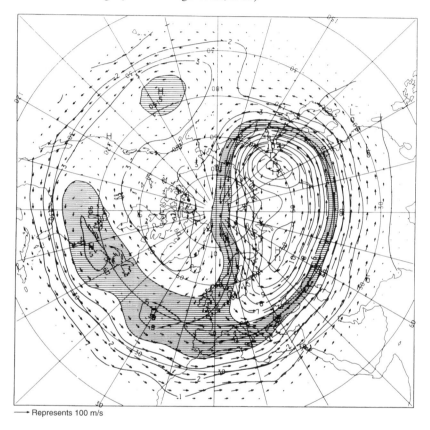

⟶ Represents 100 m/s

is a more general form of potential vorticity and is known as Ertel's potential vorticity. Equation (8.61) is Ertel's theorem. It states that providing there is negligible friction or dissipation, the Ertel potential vorticity is conserved following the fluid motion.

It is a very important theorem which forms the basis of much geophysical fluid dynamics (Gill, 1982, and Hoskins, McIntyre & Robertson, 1986). Fig. 8.8 illustrates the distribution of potential vorticity on an isentropic surface in the stratosphere where, because it is a conserved quantity, potential vorticity is a useful tracer of air motion.

Fig. 8.9. Vertical cross-section from Green Bay, Wisconsin (GRB) to Apalachicola, Florida (AQQ) for 0000 GMT on 19 February 1979 showing isentropes (solid, K), geostrophic wind (dashed, ms^{-1}). The potential vorticity (dark solid, units $10^{-5}Ks^{-1}kPa^{-1}$) analysis is shown only for the upper portion of the frontal zone and the stratosphere. (After Uccellini et al., 1985)

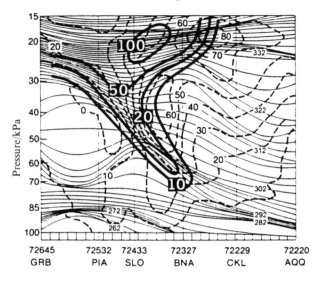

8.23 For an atmosphere at rest the Ertel potential vorticity (equation 8.61) is approximately

$$\frac{f}{\rho}\cdot\frac{\partial\theta}{\partial z}$$

For such an atmosphere plot values of the Ertel potential vorticity as a function of height and as a function of θ for a standard March atmosphere at 10N, 40N and 70N (Appendix 5), showing how its magnitude is dominated in the troposphere by the latitudinal variation of f and in the stratosphere by the large values of $\partial\theta/\partial z$. Fig. 8.9 illustrates how potential vorticity can be employed to demonstrate the intrusion of stratospheric air to lower levels in the vicinity of a frontal surface.

9

Turbulence

9.1   The Reynolds number

In writing down the equations of motion (§7.2) a term F was included to allow for friction. How to deal with this term is the subject of this chapter.

To begin with, consider fluid flow along a pipe. When all elements of fluid are moving along the direction of the fluid's mean motion the flow is *laminar* and internal friction in the fluid will be due only to molecular viscosity. When, however, individual elements are moving irregularly compared with the mean motion the flow is *turbulent*.

Reynold's classical experiments on the flow of fluid of uniform density through pipes showed that a non-dimensional quantity could be found whose approximate value determined whether the sheared flow was laminar or turbulent. This quantity is known as the Reynolds number Re; for a fluid of kinematic viscosity $\nu^{\dagger}$ moving with velocity V it is given by

$$Re = LV/\nu$$

where L is a typical length scale of the motion (e.g. the diameter of the pipe). For large values of Re greater than about 6000, turbulent flow occurs, for smaller values the flow is laminar. For air at STP $\nu \simeq 1.5 \times 10^{-5}\,\mathrm{m^2\,s^{-1}}$ so that if $L = 1\,\mathrm{m}$, provided $V > {\sim}0.1\,\mathrm{m\,s^{-1}}$ the condition that $Re > 6000$ is satisfied. With typical atmospheric scales and velocities, therefore, we expect the flow to be turbulent.

It is useful in any consideration of a particular type of atmospheric motion to restrict the consideration to a particular scale; in the last chapter this was done in order to isolate particular types of wave motion. The fact of turbulence means that this isolation can never in fact be complete. Motion on all smaller scales than the one being considered will exist;

---

† The kinematic viscosity *is the viscosity* divided by the density – its units are $\mathrm{m^2\,s^{-1}}$.

131

interactions and exchange of energy will take place between motion on these different scales. One of the main objects, therefore, of the theory of turbulence is to provide means whereby, for any particular problem, the effect of turbulent motions on smaller scales than the one being considered can be expressed in terms of the parameters of the mean flow.

In this chapter we shall first describe the effect of turbulence in the boundary layer close to the earth's surface and then discuss in rather general terms the interactions between motion on various scales in the atmosphere at large.

9.2   Reynolds stresses

At any given point and time the components of velocity u, v and w may be expressed in terms of mean components (denoted by bars) and fluctuating components (denoted by primes), e.g. $u = \bar{u} + u'$. For this to make sense, a suitable averaging period has to be found such that the mean quantity is substantially independent of the precise averaging period. This period will depend on the particular problem under consideration; for motion in the boundary layer, 5 or 10 min is a suitable value (figs. 9.1 and 9.4).

The horizontal momentum equations (7.8) neglecting friction when due only to viscosity are

$$\frac{\partial u}{\partial t} + u\frac{\partial u}{\partial x} + v\frac{\partial u}{\partial y} + w\frac{\partial u}{\partial z} - fv + \frac{1}{\rho}\frac{\partial p}{\partial x} = 0 \tag{9.1}$$

$$\frac{\partial v}{\partial t} + u\frac{\partial v}{\partial x} + v\frac{\partial v}{\partial y} + w\frac{\partial v}{\partial z} - + fu + \frac{1}{\rho}\frac{\partial p}{\partial y} = 0 \tag{9.2}$$

and the continuity equation (7.18) is

Fig. 9.1. Typical trace of wind speed taken at a height of 10 m over Lake Ontario.

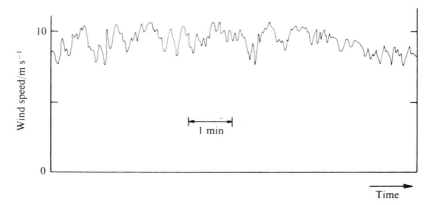

$$\frac{\partial}{\partial x}(\rho u)+\frac{\partial}{\partial y}(\rho v)+\frac{\partial}{\partial z}(\rho w)+\frac{\partial \rho}{\partial t}=0 \tag{9.3}$$

Multiplying (9.1) by ρ and (9.3) by u and adding we have

$$\frac{\partial}{\partial t}(\rho u)+\frac{\partial}{\partial x}(\rho u^2)+\frac{\partial}{\partial y}(\rho uv)+\frac{\partial}{\partial z}(\rho uw)-f\rho v+\frac{\partial p}{\partial x}=0 \tag{9.4}$$

A similar equation results from combining (9.2) and (9.3) in the same way.

We now substitute $u = \bar{u} + u'$, $v = \bar{v} + v'$ and $w = \bar{w} + w'$ into (9.4), remembering that, since $\overline{u'}, \overline{v'}, \overline{w'} = 0$ by definition, two terms only result from averaging the product of varying quantities, e.g. $\overline{uw} = \overline{(\bar{u} + u') (\bar{w} + w')} = \bar{u}\bar{w} + \overline{u'w'}$. Since relative variations in density are very much less than those in velocity, fluctuations in the density will be neglected. The resulting equation:

$$\frac{\partial \bar{u}}{\partial t}+\bar{u}\frac{\partial \bar{u}}{\partial x}+\bar{v}\frac{\partial \bar{u}}{\partial y}+\bar{w}\frac{\partial \bar{u}}{\partial z}-f\bar{v}+\frac{1}{\rho}\frac{\partial p}{\partial x}$$
$$+\frac{1}{\rho}\left[\frac{\partial}{\partial x}\overline{\rho u'u'}+\frac{\partial}{\partial y}\overline{\rho u'v'}+\frac{\partial}{\partial z}\overline{\rho u'w'}\right]=0 \tag{9.5}$$

contains all the terms of (9.1) now expressed in terms of the mean flow together with terms which express the effect of the fluctuating components; these terms describe what are known as the *Reynolds stresses* or the *eddy stresses*; they represent the flux of momentum due to the turbulent or eddy motions (problem 9.1).

The formulation of (9.5) suggests ways in which the Reynolds stresses might be measured (fig. 9.2), but gives no indication of how to express them in terms of the mean quantities. The simplest approach is to draw an analogy with molecular viscosity and considering a plane boundary in the *xy*-plane write for the eddy stress in the *x* direction on a plane parallel to the boundary

$$-\overline{\rho u'w'}=\rho K\frac{\partial \bar{u}}{\partial z} \tag{9.6}$$

where K is the *coefficient of eddy viscosity* (with the same dimensions as kinematic viscosity) and is effectively defined by (9.6). Typical atmospheric values of K lie in the range $1{-}100\,\mathrm{m^2 s^{-1}}$. These are high values when compared with the molecular viscosity of ordinary fluids (typically $10^{-5}\,\mathrm{m^2 s^{-1}}$ for gases at STP). They approximate to the viscosity of thick treacle and demonstrate the effectiveness of eddy motions compared with molecular motions in transferring momentum.

Fig. 9.2. Measurements of the three components of velocity and of the atmospheric temperature and of their fluctuations with time can be measured at different heights above the surface by a turbulence probe mounted on the tethering cable of a kite balloon. (Courtesy of the Meteorological Office Research Unit, Cardington)

9.3   Ekman's solution

Turbulent transfer is the main mechanism by which heat, moisture and momentum are exchanged between the atmosphere and the earth's surface. The region where these transfers occur is the *boundary layer*. First consider momentum transfer in this layer and let us assume a situation in which there is a shear of the mean wind in the vertical direction only. Neglecting, therefore, in (9.5) the stress involving horizontal gradients, we express the vertical eddy stress term (i.e. the third one) as in (9.6) with K as a constant. For a situation where the wind above the boundary layer is geostrophic and where horizontal temperature gradients through the boundary layer are negligible (i.e. the geostrophic wind through the layer does not change with height), the following equations describe the motion in the boundary layer

$$\frac{1}{\rho}\frac{\partial p}{\partial x} - fv - K\frac{\partial^2 u}{\partial z^2} = 0 \tag{9.7}$$

$$\frac{1}{\rho}\frac{\partial p}{\partial y} + fu - K\frac{\partial^2 v}{\partial z^2} = 0 \tag{9.8}$$

The components of the geostrophic wind (7.10) are

$$u_g = -\frac{1}{\rho f}\frac{\partial p}{\partial y}, \quad v_g = \frac{1}{\rho f}\frac{\partial p}{\partial x} \tag{9.9}$$

Substituting these in (9.7) and (9.8), multiplying (9.8) by $i = \sqrt{-1}$, and adding:

$$K\frac{\partial^2(u+iv)}{\partial z^2} - if(u+iv) + if(u_g+iv_g) = 0 \tag{9.10}$$

Suitable boundary conditions are zero wind components at the surface, i.e. $u = v = 0$ at $z = 0$ and geostrophic wind at high levels. For simplicity, assume that this geostrophic wind is zonal, i.e. $v_g = 0$, so that $u \to u_g$, $v \to 0$ as $z \to \infty$ is the other boundary condition.

The solution of (9.10) is

$$u + iv = u_g\left\{1 - \exp\left[-\left(\frac{f}{2K}\right)^{1/2}(1+i)z\right]\right\} \tag{9.11}$$

or in components

$$u = u_g[1 - \exp(-\gamma z)\cos \gamma z] \tag{9.12}$$

$$v = u_g \exp(-\gamma z)\sin \gamma z \tag{9.13}$$

where $\gamma = \left(\dfrac{f}{2K}\right)^{1/2}$

Equations (9.12) and (9.13) describe the *Ekman spiral* (fig. 9.3). Above the level $z = \pi/\gamma$ where $v = 0$ the wind is approximately geostrophic. Below this level the wind direction deviates very considerably from the geostrophic direction; at the surface for instance the deviation is 45°. The

Fig. 9.3. Wind hodograph for the lowest kilometre as measured by Dobson (1914) (dashed line) compared with Ekman spiral (full line). Figures on curves are heights in metres.

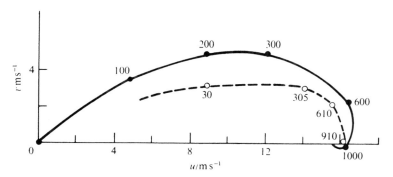

quantity π/γ may, therefore, be considered as the approximate depth of the boundary layer. With $f = 7 \times 10^{-5} \mathrm{s}^{-1}$ and $K = 10 \mathrm{m}^2 \mathrm{s}^{-1}$, $\pi/\gamma \simeq 1 \mathrm{km}$. Note that in the boundary layer the wind has a component directed generally towards low pressure, a feature which was predicted by the simple argument of §7.3.

The approximation of constant K is not a good one, particularly near $z = 0$ (cf. fig. 9.3) and also, for instance, because the boundary layer is frequently capped by an inversion below the top of the Ekman layer. Therefore the Ekman profile is not accurately followed in practice. Nevertheless as a qualitative description, the solution is a very useful one.

9.4   The mixing-length hypothesis

In the last section the eddy transfer coefficient was assumed independent of height. Especially near the surface this assumption is far from true. Some understanding of the reason for this comes from the mixing-length model of turbulence which has been valuable particularly in the near surface layers. In this model it is assumed that an element of fluid at level z moves through an average distance $|l|$ taking its velocity with it. At height $z + l$ it is reabsorbed losing all trace of its original motions. The length l therefore plays somewhat the same part as the mean free path in kinetic theory.

This description implies that

$$u' = -l\frac{\partial \bar{u}}{\partial z}$$

If the turbulence is approximately isotropic $|w'| \simeq |u'|$ and the shearing stress τ will be given by

$$\tau = -\rho\overline{u'w'} = \rho l^2 \left(\frac{\partial \bar{u}}{\partial z}\right)^2 \tag{9.14}$$

In the derivation of (9.14) it must be remembered that l is positive for upward moving parcels of fluid.

In the layer within a few tens of metres of the surface the shearing stress is approximately constant – a layer known as the *constant flux layer*. A further plausible hypothesis is that the size and path of the eddies should be proportional to height above the surface, i.e. $l = \kappa z$ where κ is known as von Karman's constant and has a value of about 0.4. On integrating (9.14) under these assumptions the wind profile is given by

$$\bar{u} = \frac{u_*}{\kappa}\ln\left(\frac{z}{z_0}\right) \tag{9.15}$$

where $u_* = (\tau/\rho)^{1/2}$ is known as the *friction velocity* and the constant of integration z_0 as the *roughness length* since it depends on the surface roughness.

Equation (9.15) fits well under conditions of neutral stability. If the stability is not neutral, buoyancy forces come into play and affect the nature of the turbulence. As might be expected, the wind profile and the associated momentum, heat and water vapour fluxes depend very considerably on the vertical stability. An important measure of the dependence of turbulence on the stability is the *Richardson number* the derivation of which is outlined in problem 9.7.

9.5 Ekman pumping

We return to the Ekman layer and assume for the sake of simplicity that the atmosphere is of uniform density of depth H, and that in the boundary layer of depth d ($d \ll H$) the wind profile is accurately described by (9.12) and (9.13), and that above the boundary layer there is a flow u_g in the x direction, independent of height but varying with the y co-ordinate. Because of friction in the boundary layer, horizontal convergence or divergence occurs which leads through the necessity for continuity to vertical motion.

The continuity equation for a situation where density changes are neglected is (7.19)

$$\frac{\partial w}{\partial z} = -\frac{\partial u}{\partial x} - \frac{\partial v}{\partial y} \tag{9.16}$$

Substituting for u and v from (9.12) and (9.13) and integrating through the boundary layer we have for the vertical velocity w_d at $z = d$:

$$w_d = -\int_0^d \frac{\partial u_g}{\partial y} \exp(-\gamma z) \sin \gamma z \, dz \tag{9.17}$$

since $\partial u/\partial x = 0$ and since on a level surface $w = 0$ at $z = 0$.

The vorticity ζ_g of the geostrophic wind above the boundary layer is equal to $-\partial u_g/\partial y$ so that on integration (9.17) becomes

$$w_d = \frac{1}{2}\zeta_g \gamma^{-1} \tag{9.18}$$

For typical values (problem 9.6), w_d is a few $\mathrm{mm\,s^{-1}}$.

This vertical velocity is a direct consequence of the existence of a surface stress and its presence is not dependent on the vertical stability.

The existence of a vertical velocity upwards from the boundary layer has consequences for the flow in the rest of the atmosphere, again

Fig. 9.4. Illustrating Ekman pumping.

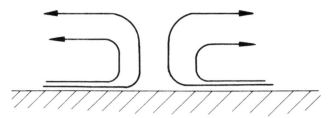

because of continuity. Suppose for instance we consider the situation in a region of cyclonic vorticity. There is inflow in the boundary layer towards the centre of the vortex, rising air above the boundary layer and a balancing outflow at higher levels (fig. 9.4). This outflow affects the vorticity ζ_g; the rate of change of ζ_g can be found from (8.36), namely

$$\frac{\partial \zeta_g}{\partial t} = f \frac{\partial w}{\partial z} \tag{9.19}$$

where ζ has been neglected compared with f in the right hand side of (8.36) (cf. problem 8.16). Integrating (9.19) between the top of the boundary layer ($z = d$) and the top of the atmosphere ($z = H$) we have

$$\frac{\partial \zeta_g}{\partial t}(H - d) = -fw_d$$

and on substituting from (9.18),

$$\frac{\partial \zeta_g}{\partial t} = -\frac{f}{2\gamma H}\zeta_g \tag{9.20}$$

since $d \ll H$.

The result expressed by (9.20) is that the vorticity is reduced with a time constant of $2\gamma H/f$ – the *spin-down time* which is typically several days. The main circulation decays very much more rapidly through this means involving a *secondary circulation* than by other damping mechanisms (problem 9.5). This secondary circulation is driven by friction in the boundary layer, a mechanism known as *Ekman pumping*.

9.6   The spectrum of atmospheric turbulence

The discussion so far in this chapter has been concerned with turbulent motion on the rather small space and time scales associated with the boundary layer. Motion on much larger scales possesses random characteristics and may also be considered as turbulence which therefore occurs in the atmosphere on a very wide range of scales. At small scales the turbulence is essentially three dimensional and not limited by the atmospheric depth, for larger scales the turbulence is approximately two dimensional. In both these regions simple dimensional arguments have been put forward in an attempt to explain the observed energy spectrum.

First suppose the existence of isotropic homogeneous three dimensional turbulence away from the atmospheric boundary. Further, suppose that in wavenumber space (wavenumber denoted by k), turbulent components at any wavenumber interact with components of similar wavenumber, and that dissipation eventually occurs at very high wavenumbers through viscosity at a rate of ε per unit mass. Under these conditions Kolmogoroff assumed that the energy spectrum $E(k)$ depends only on ε and simple dimensional considerations require (problem 9.6)

$$E(k) = \alpha_1 \varepsilon^{2/3} k^{-5/3} \tag{9.21}$$

where α_1 is a constant (cf. fig. 9.5). This is known as the *Kolmogoroff Law*.

Second, at larger scales (lower wavenumbers) the motion is quasi two dimensional. A constraint that is imposed on this quasi two dimensional motion is that the mean squared vorticity $\overline{\zeta^2}$ (the quantity $\frac{1}{2}\overline{\zeta^2}$ is called the enstrophy) must be conserved. This implies that a cascade of energy to smaller scales is no longer possible unless it is also accompanied by a much larger transfer of energy to larger scales. For two dimensional turbulence in the atmosphere (Fjortoft, 1953 and Kraichman and Montgomery, 1979) two main regimes apply depending on the main scales at which energy is inserted into the atmosphere (fig. 9.6). The first is a range where energy is transferred from small scales where energy is inserted to larger scales (a reverse energy cascade). In this range, if the energy spectrum depends only on the rate of energy dissipation, the energy spectrum of the turbulence will follow the same law as that for three dimensional turbulence described by (9.21).

The second regime for two dimensional turbulence occurs at still larger scales where it is assumed that the energy spectrum depends on the rate of transfer of enstrophy from a large scale at which there is an enstrophy source (fig. 9.6) to smaller scales. Under this assumption, a similar dimensional argument produces (problem 9.6)

$$E(k) = \alpha_2 \left(\overline{\zeta^2}\right)^{2/3} k^{-3} \tag{9.22}$$

where α_2 is a constant.

Fig. 9.5. Spectral energy density $E(k)$ of vertical velocity at different wavenumbers k, as measured at heights $z = 1$ m and $z = 4$ m under a wide range of vertical stability (different values of Richardson number Ri (cf. problem 9.7) shown against symbols). The line has a slope of $-\frac{5}{3}$. (From Matveev, 1967)

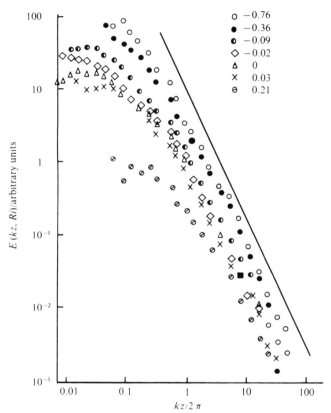

In fig. 9.7 is demonstrated the applicability of these two regimes to wind measurements made from aircraft covering a wide range of conditions and locations and horizontal scales from 20 km to 5000 km. The universality of the results suggests that energy is being injected into the atmosphere's circulation predominantly at scales of 5000 km or greater and also at scales of the order of 10 km (problem 9.8). Observations of the spectra of temperature measurements show the same regimes and slopes.

General considerations of this kind clearly do not add very much to understanding of the detail of the atmospheric circulation. Different types of motion both regular and irregular, including various kinds of waves, breaking waves, eddies, convective systems etc, all contribute to what is called turbulence. However, such considerations are of importance

Fig. 9.6. Schematic representation of spectral ranges in quasi two dimensional turbulence. (After Gage & Nastrom, 1986).

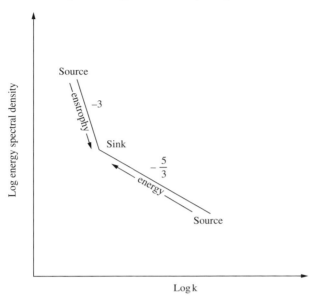

in understanding the degree of interaction between motion of different scales in space and time, which understanding has an influence on estimates of the fundamental predictability of different aspects of the atmosphere's circulation. It also influences the way in which descriptions of small scale motions are included in numerical models of the atmosphere's circulation (see chapter 11).

Problems

9.1 Show that the product $\rho u w$ is the instantaneous flux of u momentum in the z direction across an element of area parallel to the earth's surface where ρ, u and w are values appropriate to the position of the element of area. Show that when there is no mean vertical velocity the average momentum flux is $\rho \overline{u'w'}$.

9.2 If T' and m' represent respectively fluctuations in potential temperature and in water vapour mixing ratio, derive the following expressions for the vertical flux of heat Q and water vapour E due to turbulent motion

$$Q = \rho c_p \overline{w'T'}, \quad E = \rho \overline{w'm'} \tag{9.23}$$

Fig. 9.7. Wavenumber spectra of zonal and meridional velocity measured from specially instrumented Boeing 747 aircraft in routine commercial service with measurements covering all seasons and a wide range of latitudes and mostly in the altitude range 9–14 km (from Gage & Nastrom, 1986). The meridional wind spectra are shifted one decade to the right. The regimes with slopes of −3 and −5/3 are clearly demonstrated with a cross-over point between the two of about 1000 km.

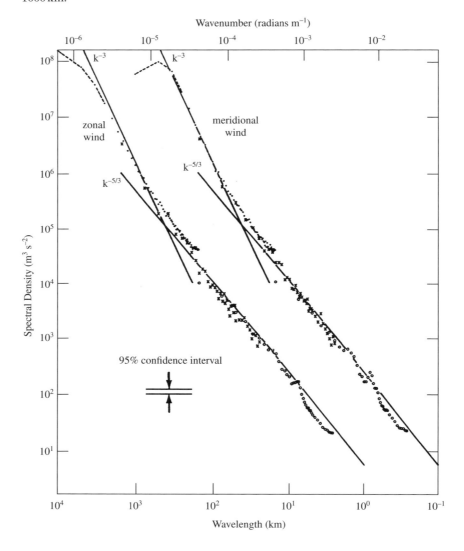

Near the surface E is also the rate of evaporation. The quantity Q/LE (where L is the latent heat of evaporation) is known as the *Bowen ratio*; it clearly depends on the moisture which is available at the surface and varies from infinity in the desert to a value of about 0.1 over the ocean.

9.3 In a similar way to (9.6), eddy transfer coefficients for heat K_Q and for water vapour K_E can be defined by

$$Q = -\rho K_Q \frac{\partial \theta}{\partial z}, \quad E = -\rho K_E \frac{\partial m}{\partial z} \qquad (9.24)$$

By considering the physical processes involved, discuss how similar the quantities K (from 9.6), K_Q and K_E may be expected to be.

9.4 In the constant flux layer the shear stress is ρu_*^2 and the heat flux Q (cf. problem 9.2). Show by dimensional analysis that a length L given by

$$L = c_p \rho T u_*^3 / \kappa g Q \qquad (9.25)$$

can be formed from the quantities where von Karman's constant κ is also included. L is known as the Monin–Oboukhov length. Similarity theory postulates that vertical gradients of heat and momentum should be functions of z/L.

9.5 Put in typical values for ζ_g, K and f in (9.18) and (9.20) and work out values of w_d and of the spin-down time. Show that the spin-down time is very much shorter than the decay time which would occur under damping due to eddy viscosity but without allowing for the possibility of vertical motion and the secondary circulation.

9.6 By dimensional analysis deduce (1) the Kolmogoroff $-\frac{5}{3}$ power law equation (9.21), (2) equation (9.22).

9.7 The nature of turbulence near the surface depends on the balance between (1) the energy which can be derived from breakdown of the mean flow and (2) the working of buoyancy forces arising from the transfer of heat away from the surface.

Neglect all horizontal gradients in applying the horizontal momentum equation (9.5) to a region of the boundary layer. Leaving the acceleration and the eddy stress terms only write it as

$$\frac{\partial}{\partial t}(\rho \bar{u}) = -\frac{\partial}{\partial z}\overline{\rho u' w'} \qquad (9.26)$$

Multiplying both sides of (9.26) by $\bar{u}$ derive the following expression for the rate of change of kinetic energy:

$$\frac{\partial}{\partial t}\left(\frac{1}{2}\rho \bar{u}^2\right) = -\frac{\partial}{\partial z}\left(\bar{u}\,\overline{\rho u' w'}\right) + \overline{\rho u' w'}\frac{\partial \bar{u}}{\partial z} \qquad (9.27)$$

Integrate (9.27) between levels z_1 and z_2 over volume V to give

$$\frac{\partial}{\partial t}\int_V \frac{1}{2}\rho\bar{u}^2 dV = -\left[\bar{u}\rho\overline{u'w'}\right]_{z_1}^{z_2} + \int_V \rho\overline{u'w'}\frac{\partial\bar{u}}{\partial z}dV \qquad (9.28)$$

The left hand side of (9.28) is the rate of change of kinetic energy in the mean flow. On the right hand side the first term is the work done due to the drag at the boundaries z_1 and z_2 and the second term is the rate of transfer of energy from turbulence to the mean flow.

Show therefore by referring to (9.14) that the rate of destruction of turbulence per unit volume by transfer to the mean flow is $-\tau\,\partial\bar{u}/\partial z$.

Now considering the effect of buoyancy forces show that the upward force on an element of unit volume at temperature $\bar{T} + T'$ compared with the surroundings at temperature $\bar{T}$ is $T'g\rho/\bar{T}$. Hence show that the rate of work done on the eddies by buoyancy forces is $Qg/\bar{T}$ per unit volume where $Q = c_p\rho\overline{w'T'}$ (cf. problem 9.2). By comparing these expressions for the rate of work done on the eddies and for the rate of destruction of turbulence show that turbulence will persist if $Ri_F < 1$ where

$$Ri_F = -\frac{g}{\bar{T}}\frac{Q}{c_p\tau\,\partial\bar{u}/\partial z} \qquad (9.29)$$

This dimensionless quantity Ri_F is known as the *flux Richardson number*. Now in (9.29) replace τ and Q by the expressions in (9.6) and (9.24) and show that turbulence will persist provided $Ri < K/K_Q$ where

$$Ri = \frac{g\,\partial\theta/\partial z}{\bar{T}(\partial\bar{u}/\partial z)^2} \qquad (9.30)$$

The quantity Ri is the *Richardson number*. Under the assumption which is often made that $K \simeq K_Q$ the condition for persistent turbulence is that $Ri < 1$.

9.8 Discuss what processes or motions might be responsible for the energy or enstrophy sources or sinks at the different scales of motion presented in figs. 9.6 and 9.7.

10

The general circulation

10.1 Laboratory experiments

In §4.11 (fig. 4.5) we noticed that the atmosphere receives an excess of net radiation in equatorial regions and a deficit in polar regions. A pattern of circulation is, therefore, set up in the atmosphere which transfers heat energy from low latitudes to high latitudes.

Some insight into the type of flow to be expected in the atmosphere of a rotating planet comes from laboratory experiments on the convection which occurs in a rotating fluid. Fig. 10.1 illustrates the dependence of the type of flow on rotation rate for a laboratory experiment in which a water–glycerol solution contained in an annulus heated on the inside and cooled on the outside is rotated at different angular speeds. At low speeds the main flow is axially symmetric and heat is transferred from the centre to the outside mainly by a simple convective cell confined to the boundary layer. As the rotation rate is increased, very little transport of heat occurs unless there are significant nonaxially symmetric motions (problem 10.2). Waves appear known as *baroclinic waves* rather similar to the Rossby waves discussed in §8.4. These waves at first are regular but at higher rotation rate the preferred wavelength becomes smaller and irregularities in the wave structure occur. The main transport of heat across the annulus takes place within these waves through the mechanism of *sloping convection* (§10.6).

Fig. 10.2 illustrates that approximations to these simple regimes occur in the atmosphere. At latitudes near the equator the main transports occur through a meridional circulation; in mid latitudes large scale eddies are the dominant means of transport of heat and momentum.

A further understanding of the mechanisms of energy transfer comes from inspecting various terms in the budget of potential and kinetic energy as described in §3.6 and fig. 3.3. Much of the zonal available

Fig. 10.1. Streak photographs illustrating the dependence of the flow type on rotation rate Ω for a laboratory 'dishpan' experiment. The values of Ω in $\mathrm{rad\,s^{-1}}$ are (a) 0.41; (b) 1.07; (c) 1.21; (d) 3.22; (e) 3.91; (f) 6.4. Working fluid was a water–glycerol solution of mean density $1.037\,\mathrm{g\,cm^{-3}}$ and kinematic viscosity $1.56 \times 10^{-2}\,\mathrm{cm^2\,s^{-1}}$. The streak photographs show the flow at a depth of 0.5 cm below the free upper surface (see also problem 10.1). (From Hide & Mason, 1975)

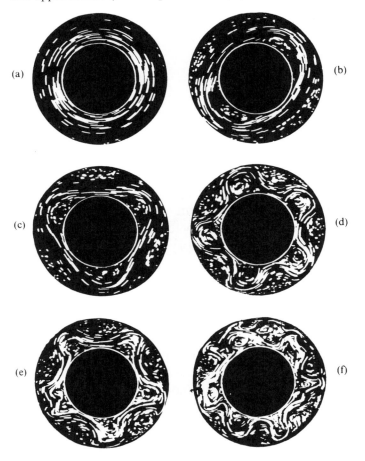

potential energy generated by the net source of radiation at the equator and net sink over the poles is converted to eddy available potential energy and eddy kinetic energy by means of the large scale eddy motions at mid latitudes.

In chapter 7, much emphasis was placed on the importance of the geostrophic approximation for describing large scale motion within the free atmosphere. It must also be realized that regions exist within the free atmosphere such as those close to fronts (cf. fig. 10.2) where strong gradi-

Fig. 10.2. Schematic features of the atmospheric circulation in winter showing Hadley circulation in tropics, sloping convection at mid and high latitudes, and the subtropical jet (STJ) associated with break in tropopause at ~30° latitude.

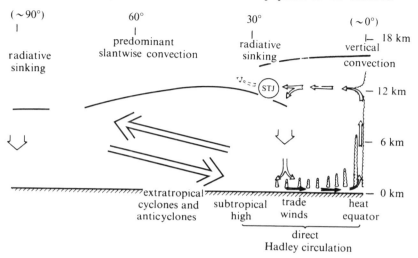

ents exist and where the flow is strongly ageostrophic. Hide (1982) has provided a general argument as to why this should be so. Comparison between the equation for the geostrophic approximation (7.10) and the full equation of motion (7.8) shows that (7.10), being lower in order than (7.8), is mathematically degenerate. Geostrophic flow therefore cannot provide a solution for all the necessary boundary conditions: the complete solution must include all the terms of (7.8) in the analysis. This implies that (Hide, 1982) regions of highly ageostrophic flow will occur not only near the boundaries of the system but also in localized regions (e.g. fronts, jet streams) of the main body of the fluid.

In the paragraphs that follow we shall look rather briefly first at a symmetric convective cell, and then at various forms of instability, especially *baroclinic instability*.

10.2 A symmetric circulation

We want first to find out whether a steady-state solution to the equations of motion can be found for an atmosphere on a rotating planet heated at low latitudes and cooled at high latitudes in which the flow shows axial symmetry, i.e. no variation with longitude.

To simplify the discussion (which partially follows Gill, 1982) the Boussinesq approximation will be assumed, i.e. changes of density are neglected except where they are coupled with gravity to produce buoyancy forces (cf. §8.6).

We shall assume that the atmosphere is heated at a rate which varies with latitude and altitude. We also need to allow for the action of radiation processes tending to restore the atmosphere to its equilibrium situation (cf. §2.4). The simplest way to do this is, in the thermodynamic equation, to replace $\partial/\partial t$ by $(\partial/\partial t) + \alpha$ where $1/\alpha$ is the time constant of the radiative processes (cf. §2.4). When these terms are added to the thermodynamic equation (8.42) remembering that no variations are occurring in the x direction, we have

$$\left(\frac{\partial}{\partial t}+\alpha\right)\left(\frac{\rho'}{\rho}\right)-\frac{N_B^2}{g}w+Q=0 \tag{10.1}$$

where Q is the heating term and ρ' is the density perturbation associated with the circulation.

The appropriate momentum equations are (8.25) and (8.26), but, because we are looking for a steady-state solution, terms describing frictional dissipation need to be added. The simplest way to do this is, in the momentum equations, to replace $\partial/\partial t$ by $\partial/\partial t + r$ where $1/r$ is the time constant of the frictional decay process. Friction of this kind which is assumed to depend linearly on the velocity is known as *Rayleigh friction*. The momentum equations therefore become (remembering that all x variations are assumed zero)

$$\left(\frac{\partial}{\partial t}+r\right)u-fv=0 \tag{10.2}$$

$$\left(\frac{\partial}{\partial t}+r\right)v+fu=-\frac{1}{\rho}\frac{\partial p'}{\partial y} \tag{10.3}$$

Also required is the hydrostatic equation (8.41), i.e.

$$\frac{1}{\bar{\rho}}\frac{\partial p'}{\partial z}+\frac{\rho'}{\bar{\rho}}g=0 \tag{10.4}$$

Looking for a steady-state solution (i.e. $\partial/\partial t = 0$) and eliminating u, ρ' and p' from equations (10.1)–(10.4), we have

$$\frac{N_B^2}{g}\frac{\partial w}{\partial y}-\frac{\alpha}{rg}\left(f^2+r^2\right)\frac{\partial v}{\partial z}-\frac{\partial Q}{\partial y}=0 \tag{10.5}$$

The equation of continuity (7.18) for the steady-state solution is

$$\frac{\partial v}{\partial y}+\frac{\partial w}{\partial z}=0 \tag{10.6}$$

Fig. 10.3. The domain for the Hadley circulation calculation.

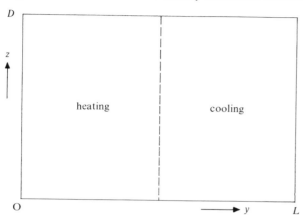

A stream function ψ can therefore be defined by the equations

$$\frac{\partial \psi}{\partial y} = w \tag{10.7}$$

$$\frac{\partial \psi}{\partial z} = -v \tag{10.8}$$

which on substituting into equation (10.5) gives the equation for ψ

$$\frac{N_B^2}{g}\frac{\partial^2 \psi}{\partial y^2} + \frac{\alpha}{rg}(f^2 + r^2)\frac{\partial^2 \psi}{\partial z^2} - \frac{\partial Q}{\partial y} = 0 \tag{10.9}$$

We consider the flow to be confined to the region $0 \leqslant y \leqslant L, 0 \leqslant z \leqslant D$ (fig. 10.3) in which, to keep the solution simple, we shall assume the Coriolis parameter f to be constant.

A reasonable simulation of the variation of heating rate with altitude and latitude results from writing Q in the form

$$Q = Q_0 \cos ly \sin mz \tag{10.10}$$

where $l = \pi/L$, and $m = \pi/D$

A rigid lid (i.e. $w = 0$) is assumed at the top of the atmosphere which, for this simple model, can be thought of as being at the tropopause. The stream function ψ can therefore be set to zero at $z = D$.

If a rigid surface is also assumed at the lower boundary, $\psi = 0$ at $z = 0$. In this case the solution of (10.8) is

$$\psi = \psi_0 \sin ly \sin mz \tag{10.11}$$

where
$$\psi_0 = \frac{lQ_0}{\dfrac{N_B^2}{g}l^2 + \dfrac{\alpha m^2}{rg}(f^2 + r^2)} \tag{10.12}$$

The solution (10.11) is plotted in fig. 10.4. Air rises in the region of maximum heating and sinks in the region of maximum cooling.

If, instead of a rigid lower boundary, it is assumed that an Ekman layer (§9.3) is present, much of the reverse flow is concentrated at low levels (fig. 10.4 and problems 10.5 and 10.6). Note also the easterly wind at low levels associated with the equatorward flow (cf. equation (10.2)). The resultant flow at low levels gives rise to the *trade winds* known by mariners for centuries (fig. 10.5). The circulation we have described is a *Hadley circulation* after Hadley who in 1735 described such a thermally driven cell and drew attention to the effect of the earth's rotation on air moving across circles of latitude.

This simple model of an axially symmetric Hadley circulation has made a number of simplifying assumptions. It has, for instance, assumed a constant value of f – a very poor assumption near the equator. Some insight into the situation at low latitudes can be obtained by considering what would happen if angular momentum were conserved as air moves northwards away from the equator. The specific angular momentum of a ring of air moving with zonal velocity u at latitude ϕ is $(u + \Omega a \cos \phi)\, a \cos \phi$. A ring of air with zero zonal velocity at the equator (and angular momentum Ωa^2), if it moves northwards while conserving its angular momentum will possess a zonal velocity at latitude ϕ of $\Omega a \sin^2 \phi / \cos \phi$. At $\phi = 15°$ this is $32\,\mathrm{m\,s^{-1}}$ and at $\phi = 30°$, $134\,\mathrm{m\,s^{-1}}$. Air that has been heated at the equator, therefore, and moves northwards in the upper troposphere can be expected to speed up rapidly because of these angular momentum considerations. In fact, the subtropical jet at around 30° latitude and at an altitude of around 12 km possesses a maximum velocity of about $40\,\mathrm{m\,s^{-1}}$. A more complete model of the Hadley circulation (see James, 1994) would take into account all sources of heating (including latent heat release) in addition to friction in the free atmosphere as well as in the boundary layer.

It is not possible for the symmetric Hadley circulation to dominate at all latitudes. We have already mentioned in §10.1 the importance of non-axially symmetric motions which, from the simple arguments presented there, would be expected to occur at higher latitudes. A further fundamental reason why other types of circulation must occur concerns the exchange of momentum between the solid earth and the oceans and the atmosphere. A Hadley circulation implies easterly flow at low levels, and

Fig. 10.4. The solution of equation (10.11) showing a Hadley circulation with rising air over the equator where the air is heated and falling air away from the equator where the air is, relatively, being cooled. The solid lines are contours of the stream function ψ with the contour interval $\frac{1}{4}$ of the maximum value. The dashed lines are contours of the x component of velocity with contour interval $\frac{1}{4}$ of the maximum speed. The dotted line is for zero zonal wind. For (a) there is no friction at the lower boundary, for (b) friction is described by problem 10.5 with the parameters ψ and λD (see problems 10.5 and 10.6) both equal to unity.

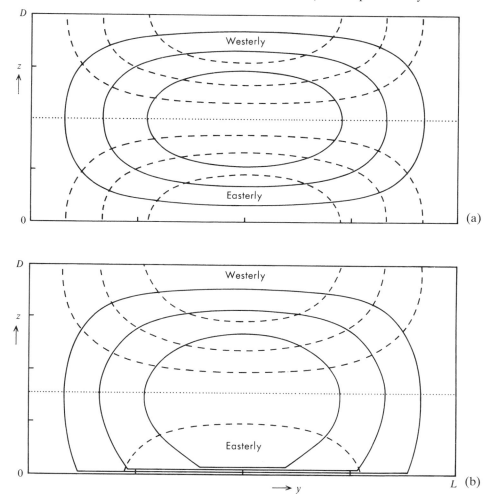

hence transfer of easterly momentum to the surface. Since on average there can be no exchange of momentum between the atmosphere and the surface, the transfer of easterly momentum in the Hadley circulation must be balanced by a transfer of westerly momentum associated with the different flow regime at higher latitudes.

Fig. 10.5. Vector mean winds over the ocean for January. (From *Meteorology for Mariners*, published by the Meteorological Office)

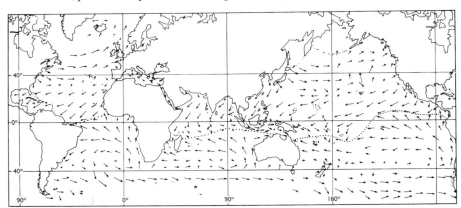

In the next few sections, we investigate the various forms of instability which occur in the atmosphere, concentrating on those that are important at high latitudes.

10.3 Inertial instability

For steady zonal flow on a rotating planet in which the geostrophic zonal velocity $\bar{u}(y)$ varies with y but not with z, consider the stability of a 'parcel' displaced with velocity v in the y direction from y_0 to $y_0 + y'$. The equations of motion describing the parcel's change in velocity are (§7.2)

$$\frac{Du}{Dt} = fv \tag{10.13}$$

and $$\frac{Dv}{Dt} = f(\bar{u} - u) \tag{10.14}$$

Since $v = dy/dt$, (10.13) may be integrated to give

$$u(y_0 + y') - \bar{u}(y_0) = fy' \tag{10.15}$$

Substituting for $u(y_0 + y')$ in (10.14) we have

$$\frac{Dv}{Dt} = \frac{d^2 y'}{dt^2} = f\left(\frac{\partial \bar{u}}{\partial y} - f\right) y' \tag{10.16}$$

There will therefore be stable oscillations in y' or exponential growth according as $(\partial \bar{u}/\partial y - f)$ is $<$ or > 0. In the northern hemisphere with positive f the condition for *inertial stability* is therefore that $f - \partial \bar{u}/dy > 0$, i.e. that the absolute vorticity of the basic flow be positive. Observations in the atmosphere show that, on the synoptic scale, this is always the case (cf. problem 10.7).

10.4 Barotropic instability

In §10.3, by the simple 'parcel' technique, a particular form of instability was investigated in a similar way to the instability with respect to vertical motion considered in §1.4. There are many other forms of instability which occur in fluid motion. In this section and the next we shall consider the stability of planetary waves which were introduced in §§8.4 and 8.6 in two particular cases. In this section we consider a *barotropic* situation, i.e. one in which there are no horizontal temperature gradients and the basic zonal flow does not vary with height. In the next section we consider a *baroclinic* situation, i.e. one in which horizontal temperature gradients occur with a resulting shear of the zonal wind with height.

For the barotropic case the basic wave equation is the statement that absolute vorticity is conserved, i.e. (8.29):

$$\frac{D(\zeta + f)}{Dt} = 0$$

In §8.4 in deriving the basic equation for Rossby waves, we assumed uniform zonal flow $\bar{u}$. Here we allow for a varying zonal flow with latitude so that we may find for what conditions of zonal flow instability may arise.

The perturbation form of (8.29) is

$$\left(\frac{\partial}{\partial t} + \bar{u}\frac{\partial}{\partial x}\right)\zeta' + v'\frac{\partial(\bar{\zeta} + f)}{\partial y} = 0 \tag{10.17}$$

where $\dfrac{D}{Dt} = \dfrac{\partial}{\partial t} + \bar{u}\dfrac{\partial}{\partial x}$

Substituting the stream function ψ defined by (8.31)

$$u' = -\frac{\partial \psi}{\partial y}, \quad v' = \frac{\partial \psi}{\partial x}$$

we have

$$\left(\frac{\partial}{\partial t} + \bar{u}\frac{\partial}{\partial x}\right)\nabla^2\psi + \frac{\partial \psi}{\partial x}\frac{\partial(\bar{\zeta} + f)}{\partial y} = 0 \tag{10.18}$$

For a wave solution

$$\psi = \operatorname{Re}\{\psi_0(y)\exp[ik(x - ct)]\} \tag{10.19}$$

where the phase velocity c may be complex, i.e.

$$c = c_r + ic_i \tag{10.20}$$

If $c_i > 0$, the wave will grow exponentially with time, i.e. it will be unstable. To find in what situation this may be the case substitute from (10.19) in (10.18) so that

$$(\bar{u} - c)\left(\frac{d^2\psi_0}{dy^2} - k^2\psi_0\right) + \psi_0 \frac{\partial(\bar{\zeta} + f)}{\partial y} = 0 \qquad (10.21)$$

To impose some boundary conditions, suppose that the wave is confined to a zonal channel of width W so that $\psi(y) = 0$ at $y = 0$ and $y = W$. There will now be solutions of (10.21) for certain values of the velocity c. Since the quantity ψ_0 is in general complex let us write it as $\psi_r + i\psi_i$. Multiplying (10.21) by the complex conjugate ψ_0^* of ψ_0, dividing by $\bar{u} - c$, integrating over y and equating real and imaginary parts, we have for the imaginary part

$$\int_0^W \left(\psi_i \frac{d^2\psi_r}{dy^2} - \psi_r \frac{d^2\psi_i}{dy^2}\right) dy = c_i \int_0^W \frac{\partial(\bar{\zeta} + f)}{\partial y} \frac{|\psi_0|^2}{|\bar{u} - c|^2} dy \qquad (10.22)$$

The integrand on the left hand side is

$$\frac{d}{dy}\left(\psi_i \frac{d\psi_r}{dy} - \psi_r \frac{d\psi_i}{dy}\right) \qquad (10.23)$$

Since $\psi_i = \psi_r = 0$ at both boundaries the integral of this expression is equal to zero. The integral on the right hand side, therefore, must be equal to zero. On inspecting the integrand we see that for $c_i > 0$, it is necessary that $\partial(\bar{\zeta} + f)/\partial y$ change sign somewhere in the region $0 < y < W$. This therefore is a necessary condition for instability. It was first worked out by Rayleigh, hence it is known as *Rayleigh's criterion*. Since $\partial\bar{\zeta}/\partial y = -\partial^2\bar{u}/\partial y^2$ and $\partial f/\partial y = \beta$ (the β-plane approximation introduced in §8.4), $\partial(\bar{\zeta} + f)/\partial y$ may be expressed as $\beta - \partial^2\bar{u}/\partial y^2$, so that the condition is that the quantity $\beta - \partial^2\bar{u}/\partial y^2$ should change sign somewhere in the region $0 < y < W$.

It is possible that this barotropic instability is responsible for the formation of tropical depressions in the *intertropical convergence zone* (ITCZ) (fig. 10.6), but it is not the form of instability responsible for the main mid-latitude disturbances.

10.5 Baroclinic instability

In the last section we considered the propagation of planetary waves under conditions of zonal flow which varied with latitude but not with altitude. Following Eady (1949) we now consider the growth of planetary waves under the same general conditions as those considered for the

Fig. 10.6. Variations of zonal wind and vorticity near the intertropical convergence zone illustrating the strong shear and the possibility of barotropic instability. (After Charney, 1973)

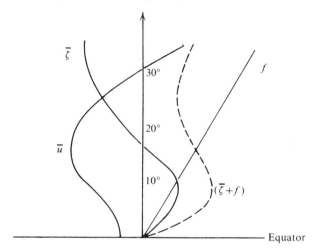

axially symmetric circulation in §10.2, i.e. in a Boussinesq fluid confined between rigid horizontal planes at $z = -H/2$ and $H/2$ with a horizontal temperature gradient such that the mean zonal flow $\bar{u}$ possesses a uniform gradient with height (cf. (7.29), (10.1) and problem 10.8)

$$\frac{\partial \bar{u}}{\partial z} = -\frac{g}{f} \frac{\partial \overline{\ln \theta}}{\partial y} \tag{10.24}$$

where θ is the potential temperature. To further simplify the problem we are going to neglect the variation of Coriolis parameter with latitude, i.e. we let $\beta = 0$.

Despite this assumption we shall find that because of the baroclinic situation (i.e. the horizontal temperature gradient) planetary wave solutions exist and further that under suitable conditions instability occurs such that the waves will grow. This *baroclinic instability* is the most important form of instability in the atmosphere as it is responsible for the mid-latitude cyclones.

The relevant equations are the vorticity equation (8.45) which with $\beta = 0$ is

$$\left(\frac{\partial}{\partial t} + \bar{u} \frac{\partial}{\partial x}\right)\left(\nabla^2 \psi + \frac{f^2}{N_B^2} \frac{\partial^2 \psi}{\partial z^2}\right) = 0 \tag{10.25}$$

and the thermodynamic equation

$$\frac{d\ln\theta}{dt} = 0$$

which on introducing the perturbations becomes

$$\left(\frac{\partial}{\partial t} + \bar{u}\frac{\partial}{\partial x}\right)(\ln\theta)' + v'\frac{\partial\overline{\ln\theta}}{\partial y} + w'\frac{\partial\overline{\ln\theta}}{\partial z} = 0 \tag{10.26}$$

The thermal wind equation (10.24) applied to the perturbation demands

$$\frac{\partial\psi}{\partial z} = \frac{g}{f}(\ln\theta)' \tag{10.27}$$

so that (10.26) becomes

$$\left(\frac{\partial}{\partial t} + \bar{u}\frac{\partial}{\partial x}\right)\frac{f}{g}\frac{\partial\psi}{\partial z} + \frac{\partial\psi}{\partial x}\frac{\partial\overline{\ln\theta}}{\partial y} + w'\frac{\partial\overline{\ln\theta}}{\partial z} = 0 \tag{10.28}$$

We look for wave-like solutions

$$\psi = \text{Re}\{\psi_0(z)\exp i(kx + ly - kct)\} \tag{10.29}$$

which on substitution into (10.25) lead to an equation for $\psi_0(z)$, i.e.

$$\frac{f^2}{N_B^2}\frac{\partial^2\psi_0}{\partial z^2} - (k^2 + l^2)\psi_0 = 0 \tag{10.30}$$

which has the solution

$$\psi_0(z) = A\sinh\alpha z/H + C\cosh\alpha z/H \tag{10.31}$$

with

$$\alpha = \frac{N_B H}{f}(k^2 + l^2)^{1/2} \tag{10.32}$$

Because of the rigid boundaries, boundary conditions are imposed $w' = 0$ at $z = H/2$ and $-H/2$ which on substitution from (10.31) into (10.28) lead to two simultaneous equations for A and C. These equations are only consistent if the determinant of the coefficients vanishes. An equation for c results namely (problem 10.9)

$$(c - u_0)^2 = \frac{(\Delta u)^2}{\alpha^2}\left(\frac{\alpha}{2} - \tanh\frac{\alpha}{2}\right)\left(\frac{\alpha}{2} - \coth\frac{\alpha}{2}\right) \tag{10.33}$$

where u_0 is the value of u at $z = 0$ and Δu is the difference in u between the two boundaries. The first bracket on the right hand side is always positive, the second is negative for $\alpha < 2.4$.

Fig. 10.7. Variation of growth rate kc_i of Eady baroclinic waves as a function of the parameter α from (10.33).

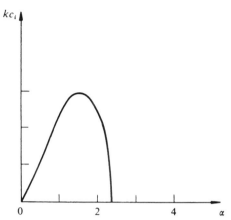

When the right hand side of (10.33) is negative, $c - \bar{u}_0$ is wholly imaginary ($= c_i$). Since the equation for the stream function contains a term $\exp(kc_i t)$, under these conditions the wave amplitude will grow exponentially with a time constant $(kc_i)^{-1}$. Note that the real part of c is equal to $\bar{u}_0$ so that the phase velocity of the growing waves is equal to that of the middle of the fluid, i.e. where $z = 0$ (sometimes known as the 'steering' level).

The maximum value of kc_i occurs when $l = 0$ and $\alpha = 1.61$ (problem 10.10 and fig. 10.7). From (10.32) for $l = 0$ the wavelength $\lambda_m = 2\pi/k$ of most rapid growth and hence of maximum instability is given by

$$\lambda_m = \frac{2\pi N_B H}{1.61 f} \qquad (10.34)$$

which for typical atmospheric values (problem 10.11) leads to a wavelength of ~4000 km. We should expect disturbances of about this wavelength to be the most common occurring in the atmosphere; a value which fits in well with typical wavelengths for disturbances found in mid latitudes. We further note that mid-latitude cyclones are observed to develop in periods of one to three days which agrees well with typical growth times found from the Eady theory (cf. problem 10.12).

The relative phases and amplitudes of pressure and temperature in the disturbance and also of the vertical and northward velocities are shown in fig. 10.8 (cf. problem 10.13).

Fig. 10. 8. Illustrating the structure of the most unstable Eady wave showing the flow at the upper surface (a) and the lower surface (b). The solid lines are isobars and the dashed lines isotherms; the phase shift between the two is exaggerated for clarity. In (a) the regions of upward ($\oplus$) and downward ($\ominus$) vertical motion at an intermediate level are also indicated. At all levels, air going poleward is generally warmer than air going equatorward, so there is a net poleward heat flux. (After Gill, 1982).

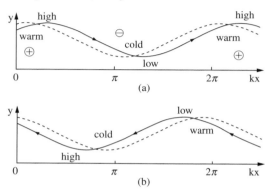

(a)

(b)

Fig. 10.9. Illustrating sloping convection.

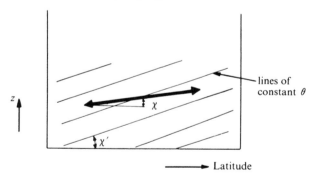

10.6   Sloping convection

The baroclinic wave described in the last section is an example of *sloping convection*. Since in the wave, northward moving air is also moving upwards and southward moving air downwards (problem 10.14), the disturbance may be regarded as an eddy in which air parcels are interchanged along a slant path of average slope χ (fig. 10.9). If $\chi < \chi'$ where χ' is the slope of the isentropic surfaces in the undisturbed state, the average potential energy will decrease and kinetic energy will be released into the eddy. If $\chi > \chi'$ eddy kinetic energy will be converted to potential energy. In a

Fig. 10.10. Northward flux of heat in northern hemisphere winter by (a) mean circulations such as the tropical Hadley cell, (b) eddies. Curve (c) is the total flux. (After Newton, 1970)

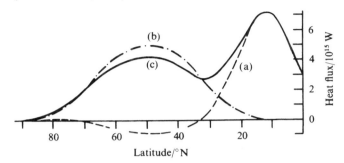

rotating system where Coriolis forces dominate over the other inertial or viscous forces, sloping convection is a more effective way of transferring heat energy than axisymmetric motions of the kind considered in §10.2. Transfer of energy by sloping convection dominates over transfer by mean meridional circulations everywhere in the atmosphere outside the Hadley circulation region in the tropics.

10.7 Energy transport

For all seasons there is an excess of net radiation over the tropics and a deficit in polar regions (fig. 4.5). As a result of the atmosphere's circulation heat is transferred from equator to poles. For any given column of atmosphere (including the surface) the energy balance may be divided into a number of terms so that

net radiation = (evaporation − precipitation) × latent heat

+ storage + net transport by atmosphere and ocean

The net radiation is the difference between incoming solar radiation and outgoing long-wave radiation, as illustrated in fig. 4.5. Storage can be in atmosphere, land or ocean; averaged over one year the net storage will be approximately zero. Transport is either by atmospheric motions or ocean currents. Fig. 10.10 shows the northward flux of heat in the northern winter by meridional motions and by eddy transfer showing that the former are effective in the tropics and the latter at higher latitudes. Heat flux due to ocean currents is difficult to estimate; it is probably comparable to that due to the atmosphere. Fig. 10.11 shows estimates of various components of the energy budget for different parts of the atmosphere, again in northern winter.

Fig. 10.11. Heat and angular momentum budgets in northern hemisphere winter. In (a), units are averages in W m⁻² for whole tropical region 5°S–32°N, and for whole extratropical region. Interior boxes give net atmospheric radiation R_a and condensation heat release LC in layers above and below 50 kPa. Dashed arrows, eddy flux e; solid arrows, cell flux c; Q_S, sensible heat flux from earth's surface. For comparison, evaporative flux Q_E is shown in parentheses. (b) shows transports of angular momentum across boundaries, and surface torques. (From Newton, 1970; for more recent data see Oort & Piexoto, 1983)

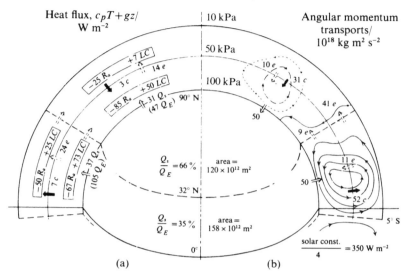

(a)   (b)

Fig. 3.3 shows more detail of the main route of the energy trans-formations. Zonal available potential energy, generated mainly by the gross differences between incoming and outgoing radiation, is converted through the eddy motions to eddy available potential energy which in turn is converted into eddy kinetic energy which is then dissipated.

In the troposphere heat is transported by atmospheric motions down the temperature gradient from the net heat source near the equator to the heat sinks around the poles. In the lower stratosphere the eddy transport of heat is still poleward but reference to fig. 5.1 will show that, because the equatorial stratosphere is colder than the polar stratosphere, this transfer is now against the temperature gradient. This can only happen if the poleward moving air parcels sink and therefore heat adiabatically while the equatorward moving ones rise and therefore cool. Such heat transport is characteristic of a refrigerator rather than a direct heat engine, the stratospheric refrigerator being driven by energy released by the tro-

Fig. 10.12. Average storages and main conversions of available potential energy E_A and kinetic energy E_K for 5–1 kPa layer in stratosphere as estimated by Crane (1977) from satellite data for period 28 January–10 February 1973. Units of energy in kJ m^{-2} and of conversion in W m^{-2}. Subscripts Z and E are for zonal and eddy components respectively. Compare with fig. 3.3 for the whole atmosphere.

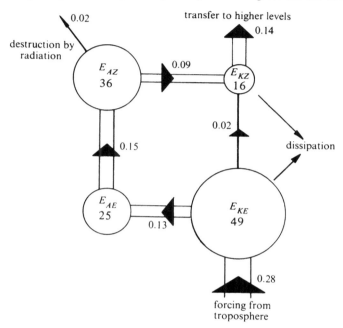

pospheric heat engine. A diagram similar to fig. 3.3 but for the stratosphere is shown in fig. 10.12, which illustrates energy transformations from the eddy kinetic energy supplied by tropospheric forcing.

10.8 Transport of angular momentum

Apart from small effects due to tidal friction the angular momentum of the earth and its atmosphere remains constant. Since the earth's average angular velocity is very close to being constant, the atmosphere's angular momentum must also be substantially constant (problem 10.16). This means that the average transfer of angular momentum between earth and atmosphere must be close to zero. A poleward transfer of angular momentum must, therefore, occur between the tropical regions where the surface winds are easterly to mid latitudes where they are westerly. In the tropical regions this transfer of angular momentum occurs both through the Hadley circulation and through eddy transfer. In mid latitudes the

Fig. 10.13. (a) Mean angular velocity about polar axis of atmosphere relative to earth's surface; and mean annual northward angular momentum transport by (b) mean circulation, and (c) by eddies. Note the up-gradient transport of angular momentum by eddies. (After Starr, 1968)

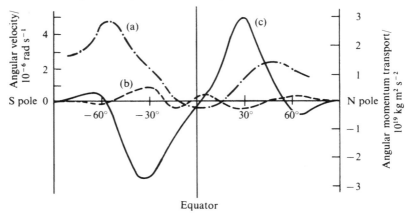

transfer occurs almost entirely through the large scale eddies (figs. 10.11 and 10.13). the torque on the surface is partially due to friction in the boundary layer and partially due to systematic pressure differences between the western and eastern sides of mountain ranges especially in the northern hemisphere.

For poleward transport of angular momentum in the eddies the quantity $\overline{u'v'}$ must be positive, i.e. there must be a significant correlation between positive values of u' and v'. In a symmetrical wave, for instance the fastest growing Eady wave of §10.5 (problem 10.17), this is not the case and the net transport will be zero. If, however, the troughs and ridges in the wave are tilted from south-west to south-east (fig. 10.14) then $u' > 0$ when $v' > 0$ and northward eddy transfer of angular momentum will occur.

Notice from fig. 10.13 that much of the zonal momentum transport by eddies occurs against the gradient, thus exhibiting an effective viscosity which is negative. Similar *negative viscosity* situations in which the momentum of eddies is fed into zonal flow have been observed in atmospheres other than that of the earth, for instance in the atmospheres of the sun (Starr, 1968), Mars (Leovy, 1969 and fig. 10.15) and Jupiter (Hide, 1974).

10.9   The general circulation of the middle atmosphere

So far in this chapter we have mostly concentrated on the dynamic processes which dominate the circulation of the lower atmosphere. We now

Fig. 10.14. Schematic representation of planetary waves in tropospheric flow. Note the tilting of the waves required to provide a northward momentum flux. Note also the depressions (L) and anticyclones (H) associated with the mean upper tropospheric flow (full line). (After Mintz, 1961)

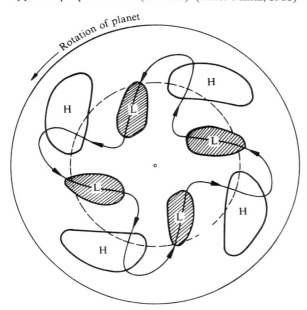

consider the circulation of the middle atmosphere – the region between about 10 and 100 km in altitude. It was noted in §10.7 and fig. 10.12 that transport in this region is largely driven from the troposphere below. Some different considerations apply, therefore, for the dynamics of the middle atmosphere.

Let us begin by asking what the temperature structure of the middle atmosphere would be if it were in radiative equilibrium. Fig. 10.16 shows this; it should be compared with fig. 5.1 where the actual temperature structure is plotted. As might be expected, the winter stratosphere and mesosphere are very much warmer than would occur under radiative equilibrium; the summer mesopause, on the other hand is very much colder.

The variation of temperature at the mesopause is just the reverse of what would be expected on grounds of radiative equilibrium. This is accomplished through an interhemispheric circulation (fig. 10.17) first suggested by Murgatroyd and Singleton (1961) and Leovy (1964). Rising air in the summer mesosphere is cooled as it ascends and is balanced by descending air in the winter mesosphere with its associated warming

Fig. 10.15. A spiral cloud formation seen in the atmosphere of the planet Mars taken from the Viking orbiter in 1977. The circulation which is about 200 km across is similar to that found in extra-tropical cyclones on earth. Temperatures measured simultaneously suggest that the cloud is formed of water ice. (Courtesy of NASA)

(problem 10.18). The upper part of the cell is completed by flow between the hemispheres in the region of the mesopause (fig. 10.17). It is easy to see in general terms how such a circulation can be associated with the thermal imbalance near the stratopause level. It is not so easy to see however what mechanism acts as the brake on the circulation. A clue to this comes from inspection of the mean zonal wind pattern shown in fig. 5.1(b). It will be noticed that the magnitude of the zonal wind relative to the earth's surface is about zero near 90 km – the mesopause level. Exchange of momentum between the surface and the mesopause which occurs through breaking gravity waves (fig. 10.18 and

Fig. 10.16. Cross-section of the atmosphere showing radiative equilibrium temperatures for the solstices in the stratosphere and mesosphere and convective conditions below. (After Murgatroyd, 1969)

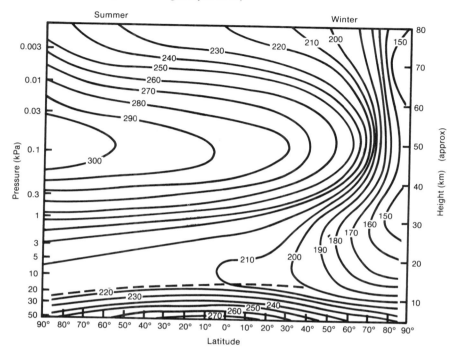

Fig. 10.17. Illustrating the inter-hemispheric circulation in the mesosphere.

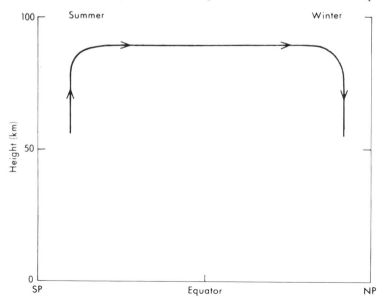

Fig. 10.18. Illustrating the behaviour of a gravity wave as it propagates upward into the mesosphere. When at level z_s the amplitude of the horizontal velocity of the wave reaches $|\bar{u} - c|$, c being the phase velocity of the wave and $\bar{u}$ the velocity of the mean flow, saturation of the wave is said to occur and momentum transfer from the wave to the mean flow begins to take place, until the critical level z_c is reached where $\bar{u} - c = 0$ (see problem 10.19). (After Fritts, 1984)

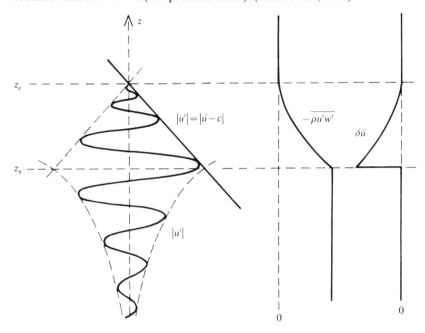

problem 10.19) operates in such a way so as to maintain this region of very small zonal flow (Houghton, 1978). The overall structure of the meso-sphere, therefore, is determined by radiative exchange together with a circulation which, in turn, satisfies (i) the thermodynamic equation, (ii) the thermal wind relationship (cf. §7.6 and fig. 5.1(a&b)), and (iii) the requirement for very small zonal flow in the region of the mesopause (problem 10.20).

Turning now to the stratosphere we note that while the summer stratosphere is not very far removed from radiative equilibrium, the winter stratosphere is very much warmer than it would be under the assumption of radiative equilibrium. We also note that the stratosphere throughout possesses a highly stable stratification with values of potential temperature far higher than is typical in the troposphere. Vertical motion is therefore severely inhibited (problem 10.18). Clearly, however, in the winter hemisphere heat is being transported from low to high latitudes by dynamical processes.

Further clues regarding the nature of the circulation come from studies of the movement of minor constituents which act as tracers. Two of these were mentioned in §5.6, namely water vapour and ozone. Very dry air which has entered the stratosphere in the region of the tropical tropopause (problem 5.12) is carried polewards through the rest of the stratosphere, all stratospheric air being found to have a water vapour content appropriate to saturated air at a temperature near to that of the tropical tropopause. Ozone-rich air from the tropical stratosphere is also, especially in the winter, carried polewards and downwards. A simple meridional circulation cell (Brewer, 1949) could explain this transport. Such a circulation by itself would not, however, conserve momentum nor would it necessarily transport an adequate amount of heat during the winter months.

Examples of planetary waves in the stratosphere were presented in §8.6. There it was pointed out that planetary waves are inhibited from propagating upwards during the regime of easterly flow which occurs in the summer, but that waves of low wavenumber (i.e. large wavelength) readily propagate upwards from the troposphere into the stratosphere during the westerly flow in the winter.

The presence of a wave does not, however, imply that transport is occurring. A steady wave motion moves air about, but unless there are dissipative processes occurring or unless the amplitude of the wave is changing with time, the air will on average be unaltered by the wave motion. Dissipation occurs through radiative damping or through interaction with smaller scale motions; an example of a large 'breaking' wave where the latter are particularly important is shown in fig. 8.8. It is these dissipative and transient processes which enable the wave to interact with the mean motions and which control the transport of heat, material and momentum by the wave (Plumb, 1982, World Meteorological Organization, 1986; Andrews, Holton & Leovy, 1987).

Measurements such as those made from Nimbus satellites and to be made from the Upper Atmospheric Research Satellite (§12.8), together with studies employing two dimensional and three dimensional models (chapter 11) are providing the means to elucidate the complex interactions between waves and mean flow and between the processes determining the radiation transfer, the chemical transformations and the dynamics – research in the whole area being organized through the Middle Atmosphere Programme (MAP).

Problems

10.1 Consider an annulus rotating at angular velocity Ω containing
liquid as shown in fig. 10.1. The equation of state relating density ρ
to temperature T for the liquid is

$$\rho = \rho_0[1 - \alpha(T - T_0)]$$

where α is the thermal expansion coefficient ($\simeq 2 \times 10^{-4}\,\mathrm{K}^{-1}$). Derive
the thermal wind relationship (cf. §7.6 and problem 7.18) for the
annulus in the form

$$\frac{\partial \mathbf{V}}{\partial z} = \frac{\alpha g}{2\Omega} \mathbf{k} \wedge \nabla T \tag{10.35}$$

where $\mathbf{V}$ is the fluid's velocity.

   If a Rossby number is defined for the annulus (whose interior and exterior radii are a and b respectively) by $R_0 = u/\Omega(b - a)$,
where u is a typical relative zonal velocity of the fluid, show by substituting in (10.35) that

$$R_0 \simeq \frac{\alpha g H \Delta T}{2\Omega^2(b-a)^2} \tag{10.36}$$

where H is the depth of the fluid and ΔT the difference in temperature across it. Estimate Ro for the situation of fig. 10.1 (a), (c) and
(e) respectively for which $\Delta T = 9\,\mathrm{K}$, $b - a = 5\,\mathrm{cm}$, $H = 14\,\mathrm{cm}$ and Ω
= 0.41, 1.21 and 3.91 $\mathrm{rad\,s}^{-1}$ respectively.

10.2 For a laboratory system of the kind described in §10.1 with axially
symmetric boundary conditions, with cylindrical co-ordinates r, ϕ,
z and with the angular velocity of rotation being parallel, to the z
axis, show that the radial velocity u_r can be written

$$u_r = -\frac{1}{2\rho\Omega_r}\left(\frac{\partial p}{\partial \phi}\right) + A \tag{10.37}$$

where ρ is the density, p the pressure and where the first term on
the right hand side is the geostrophic term; A is the sum of all the
ageostrophic terms.

   Hence show that the rate of transport by advection of any
quantity Q (per unit volume), such as heat, across a cylindrical
surface of radius r, is given by

$$H = \int_{z_1}^{z_2} \int_0^{2\pi} u_r Q r d\phi dz$$

$$= \int_{z_1}^{z_2} \int_0^{2\pi} \left[-\frac{1}{2\rho\Omega} \left(\frac{\partial p}{\partial \phi} \right) + rA \right] Q d\phi dz \qquad (10.38)$$

Experiments show that the contribution A to equation (10.38) decreases rapidly with increasing Ω. The transport H therefore will be very small unless the flow pattern departs significantly from axial symmetry. It would be expected that the flow would arrange itself in such a way that the transport of heat, for instance, would be nearer to a maximum rather than to a minimum. Even, therefore, when the boundary conditions are symmetric, departures from axial symmetric flow are to be expected (for further details of the argument see Hide, 1982).

10.3 Derive equations (10.5) and (10.9).

10.4 Show that (10.11) is a solution of (10.9).

10.5 Consider in the simple model circulation of §10.2 the presence of a boundary layer such that the lower boundary condition for the vertical velocity w is given by equation (9.18). Show now that equation (10.9) and the boundary conditions can be satisfied by the solution for the stream function

$$\psi = \psi_0 \sin ly [\sin mz + \gamma \sinh \lambda(D - z)] \qquad (10.39)$$

where $\quad \lambda^2 = \dfrac{l^2 gB}{\alpha(f^2 + r^2)} \qquad (10.40)$

and where $\quad \gamma = \dfrac{m}{\beta \sinh \gamma D + \lambda \cosh \lambda D} \qquad (10.41)$

with $\quad \beta = r \left(\dfrac{2}{Kf} \right)^{1/2} \qquad (10.42)$

where K is defined in equation (9.6).

10.6 Insert in equations (10.40) and (10.41) values of the parameters appropriate to a Hadley circulation at tropical latitudes. Show that the values of $\gamma = 1$ and $\lambda D = 1$ used to prepare fig. 10.4 are reasonable ones.

10.7 Devise a field of zonal flow for which the absolute vorticity is <0.

10.8 Write (7.29) in terms of potential temperature and hence deduce (10.24).

10.9 Derive (10.33).

10.10 From (10.33) show that the maximum value of α for which $c_i \geqslant 0$ is the solution to the equation

$$\frac{x}{2} = \coth \frac{x}{2}$$

10.11 Insert typical values into (10.34) and deduce values for λ_m.

10.12 With values of λ_m as found from problem 10.11, with $\alpha = 1.61$ and with a typical value of Δu, from (10.33) find values for $(kc_i)^{-1}$, the time constant of the rate of growth of the disturbances of greatest instability.

10.13 Substitute from (10.31) into (10.28) putting in the boundary condition $w' = 0$ at $z = \pm H/2$. Hence find the ratio C/A and show that for the wave of maximum growth the ratio is imaginary. Hence check the relative phases of quantities in the wave shown in fig. 10.8.

10.14 Show that, in the baroclinic wave of §10.5, rising air is moving northwards and sinking air southwards.

10.15 Air from the equator is moved slowly northwards while its angular momentum is conserved. What zonal velocity will it have at latitude 30°N relative to the earth's surface?

10.16 Consider the atmosphere at rest relative to the earth. What is the ratio of its angular momentum to the angular momentum of the solid earth (assume of mean density $5500\,\mathrm{kg\,m^{-3}}$)? Assuming constant zonal velocity for the atmosphere what variation in zonal velocity is required to produce the observed seasonal variation of 2 parts in 10^8 in the earth's rotation rate?

10.17 Show that for the fastest growing Eady wave of §10.5, since $l = 0$, $\overline{u'v'} = 0$, that the wave is symmetrical and no north–south momentum transport occurs.

10.18 Work out values of the potential temperature at 10 km intervals at altitudes between 10 km and 50 km for the stratosphere at 40°N in March (cf. appendix 5).

Suppose stratospheric air is cooling by radiation at a rate of $0.1\,\mathrm{K\,day^{-1}}$. How long will it take for the air to fall by 2 km?

10.19 Refer to fig. 10.18 where the behaviour of a gravity wave is illustrated as it propagates upwards into the mesosphere. We want to calculate the vertical flux of horizontal momentum. From equation

(9.5), for a basically isothermal atmosphere with mean flow $\bar{u}(z)$ this flux is given by

$$\bar{\rho}\frac{\partial \bar{u}}{\partial t} = -\frac{\partial}{\partial z}\overline{\rho u'w'} \tag{10.43}$$

Consider a gravity wave with velocities varying as (cf. §8.3)

$$u' = \text{Re}\left\{\hat{u}(z)\exp\left[\frac{z}{2H} + i(\omega t + kx + mz)\right]\right\} \tag{10.44}$$

with a similar expression for w'. The wave is said to be *saturated* when

$$\hat{u}\exp\left(\frac{z}{2H}\right) = |\bar{u} - c| \tag{10.45}$$

The result of problem 8.21 shows that, provided $m \gg 1/2H$, the vertical perturbation velocity amplitude satisfies

$$\hat{w} = -\frac{k\hat{u}}{m} \tag{10.46}$$

From equation (8.23) show that, m provided $k \ll m$,

$$m = \frac{N_B}{(\bar{u} - c)} \tag{10.47}$$

Using (10.47) and substituting for $\hat{u}$ and $\hat{w}$ in equation (10.43) show now that in a region where $\bar{u}$ varies substantially more slowly with height than does the density,

$$\frac{\partial \bar{u}}{\partial t} = -\frac{k(\bar{u} - c)^3}{2N_B H} \tag{10.48}$$

Note that the sign of the momentum exchange is such that the tendency is always to accelerate the mean flow towards the phase speed of the gravity wave.

10.20 A simple model of the interhemispheric circulation in the mesosphere, illustrated in fig. 10.17, can be set up as follows. Take the region of the mesosphere with $z = 0$ at an altitude of $50\,\text{km}$ and $z = D$ at $85\,\text{km}$ altitude. Take $y = 0$ at the equator and $y = \pi L$ at the pole. At the base of the region assume the zonal wind $u = -u_0 \sin(y/L)$ (approximates to the summer hemisphere of fig. 5.1(b)ii). Assume no variation of any quantities with x. At the top of the region near the mesopause, the boundary conditions are that u is constrained to be zero by gravity wave breaking (problem 10.19).

This means that an appropriate meridional velocity at the mesopause can be chosen to satisfy continuity. Within the region assume that the thermal wind equation (7.29) applies, i.e.

$$\frac{\partial u}{\partial z} = -\beta \frac{\partial T}{\partial y} \qquad (10.49)$$

where β will be considered constant. Also assume that away from the boundaries the meridional velocity $v = 0$, so that to satisfy continuity the vertical velocity must be given by

$$w = w_0(y)\exp(z/H) \qquad (10.50)$$

where H is the scale height which for this simple model is assumed constant. Assume the y-dependence of the temperature is governed solely by the adiabatic heating or cooling due to vertical motion, i.e. that

$$T - T_0 = -\alpha w \qquad (10.51)$$

T_0 is a mean temperature which can vary with z but not y and α is assumed independent of y or z.

Derive expressions for the variation of u and w with y and z. Note that an exactly matching circulation can be set up in the winter hemisphere. Estimate suitable values for the constants H, β and u_0, hence sketch out the fields of w, u and T.

11

Numerical modelling

11.1 A barotropic model

In previous chapters the basic equations describing atmospheric structure and motion have been solved for particular situations in order to isolate a number of specific phenomena. In the general case we need to integrate the basic equations with respect to time starting with a given atmospheric situation at a particular time so that a simulation can be provided of the atmospheric behaviour at subsequent times. The task of writing the equations and the boundary conditions in a suitable form and then of solving them with high speed digital computers is known as *numerical modelling*. By comparing the behaviour of the model with that of the real atmosphere the validity of the procedures employed by the model are tested. The most important application of numerical modelling is the development of methods sufficiently reliable and sufficiently fast to be used in routine weather forecasting. The first attempt to build such a model was made by L. F. Richardson in 1918; his book (see bibliography) is still a classic in the field. The first successful forecast was made by Charney, Fjörtoft and von Neumann (1950).

The equation employed for this early forecast was the vorticity equation applied to a barotropic atmosphere, i.e. an atmosphere which is assumed to be homogeneous and of uniform density and in which vertical motion is ignored. Writing the components of the velocity $\mathbf{V}$ in terms of the stream function ψ as defined by (8.31), the vorticity equation (8.29) becomes

$$\frac{\partial}{\partial t}(\nabla^2\psi)+\mathbf{V}\cdot\boldsymbol{\nabla}(\nabla^2\psi+f)=0 \tag{11.1}$$

Equation (11.1) is a one parameter equation which may be integrated numerically as it stands. It will apply to a level in a real atmosphere if a level can be found where the horizontal divergence of the flow is negligible and where coupling with other levels or with the boundaries may also be neglected. Synoptic flow in the mid troposphere is sufficiently non-divergent that for short periods (11.1) may be employed with some success.

11.2 Baroclinic models

With a barotropic model it is not possible to predict the development of instabilities due to thermal gradients such as were described in §10.5. For these to be present in the model we must describe the motion at more than one atmospheric level, to allow for vertical motion and to use the thermodynamic equation.

Under the assumptions that the synoptic scale motions to be described are quasi-horizontal and quasi-geostrophic, a suitable form of the vorticity equation is (8.35) which on using the continuity equation (7.20) and the stream function as defined by (8.31) can be written in isobaric co-ordinates as

$$\frac{\partial}{\partial t}(\nabla^2 \psi) + \mathbf{V} \cdot \mathbf{\nabla}(\nabla^2 \psi + f) - f\frac{\partial \omega}{\partial p} = 0 \tag{11.2}$$

where in writing the last term ζ has been neglected in comparison with f (problem 8.16).

The thermodynamic equation in isobaric co-ordinates is (using (7.34)):

$$\frac{\partial \ln \theta}{\partial t} + u\frac{\partial \ln \theta}{\partial x} + v\frac{\partial \ln \theta}{\partial y} + \omega\frac{\partial \ln \theta}{\partial p} = \frac{1}{c_p}\frac{dS}{dt} \tag{11.3}$$

where dS/dt is the rate of increase of entropy due to diabatic processes (§11.7ff).

Now

$$d \ln \theta = \frac{1}{\gamma}d \ln p + d \ln \rho^{-1} \tag{11.4}$$

from (8.15), and the hydrostatic equation (1.2) when written in terms of the geopotential Φ (problem 7.11) gives

$$\rho^{-1} = -\partial \Phi / \partial p \tag{11.5}$$

Since the first three terms of (11.3) involve differentiation at constant pressure, with the help of (11.4) and substituting from (11.5), (11.3) becomes

$$\frac{\partial}{\partial t}\left(-\frac{\partial \Phi}{\partial p}\right) + u\frac{\partial}{\partial x}\left(-\frac{\partial \Phi}{\partial p}\right) + v\frac{\partial}{\partial y}\left(-\frac{\partial \Phi}{\partial p}\right) - \frac{\omega B}{g\rho^2} = \frac{1}{\rho c_p}\frac{dS}{dt} \tag{11.6}$$

where the static stability parameter

$$B = \frac{\partial \ln \theta}{\partial z} = -g\rho \frac{\partial \ln \theta}{\partial p} \tag{11.7}$$

The geostrophic approximation (7.14) enables the geostrophic stream function ψ defined by (8.31) to be identified with Φ/f (cf. (8.44)) so that (11.6) may be written as

$$\frac{\partial}{\partial t}\left(\frac{\partial \psi}{\partial p}\right) + \mathbf{V} \cdot \nabla\left(\frac{\partial \psi}{\partial p}\right) + \frac{\omega B}{fg\bar{\rho}^2} = -\frac{1}{\bar{\rho}c_p f}\frac{dS}{dt} \tag{11.8}$$

where in the heating term and the expression for static stability an average density $\bar{\rho}$ has been included.

The vorticity equation (11.2) and the thermodynamic equation (11.8) are the basic equations for numerical integration. To see how this is carried out for a two-level model consider the atmosphere divided up into layers as shown in fig. 11.1 with the levels denoted by subscripts 0 to 4. The equations are solved for stream functions ψ_1 and ψ_3 at the levels 1 and 3. At the top of the atmosphere the boundary condition is $\omega_0 = 0$ and similarly in the absence of orograpy $\omega_4 = 0$ at the bottom. In terms of finite differences

$$\left(\frac{\partial \omega}{\partial p}\right)_1 \simeq \frac{\omega_2}{\Delta p} \quad \text{and} \quad \left(\frac{\partial \omega}{\partial p}\right)_3 \simeq -\frac{\omega_2}{\Delta p} \tag{11.9}$$

where Δp is a pressure difference of half an atmosphere. Vorticity equations as (11.2) may, therefore, be written for levels 1 and 3, namely

$$\frac{\partial}{\partial t}\nabla^2\psi_1 + \mathbf{V}_1 \cdot \nabla(\nabla^2\psi_1 + f) - f\omega_2/\Delta p = 0$$

$$\frac{\partial}{\partial t}\nabla^2\psi_3 + \mathbf{V}_3 \cdot \nabla(\nabla^2\psi_3 + f) - f\omega_2/\Delta p = 0 \tag{11.10}$$

where $\mathbf{V}_1 = \mathbf{k} \wedge \nabla\psi_1$ and $\mathbf{V}_3 = \mathbf{k} \wedge \nabla\psi_3$

Fig. 11.1. Arrangement of levels and variables for a two-parameter baroclinic model.

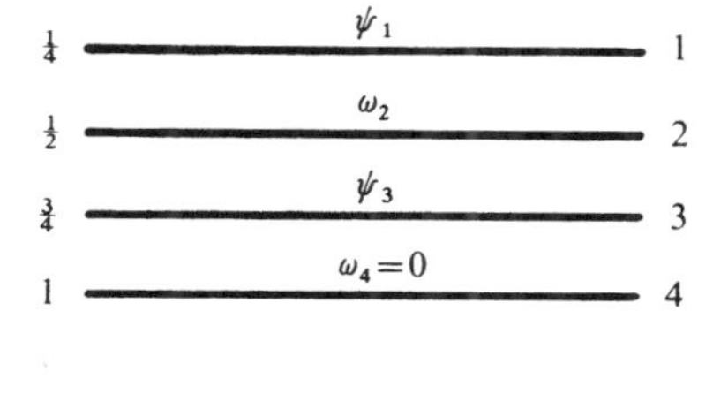

The finite difference form of (11.8) in terms of ψ_1 and ψ_3 and ω_2 only will be

$$\frac{\partial}{\partial t}(\psi_1 - \psi_3) + \mathbf{V}_2 \cdot \mathbf{\nabla}(\psi_1 - \psi_3) - \frac{B\Delta p}{gf\bar{\rho}^2}\omega_2 = \frac{1}{c_p}\frac{\Delta p}{\bar{\rho}f}\frac{dS}{dt} \qquad (11.11)$$

where $\mathbf{V}_2 = \mathbf{k} \wedge \mathbf{\nabla}\frac{1}{2}(\psi_1 + \psi_3)$

If ω_2 is eliminated between (11.10) and (11.11), two equations for ψ_1 and ψ_3 result which may be integrated numerically.

11.3   Primitive equation models

The baroclinic model described in §11.2 although less restrictive than the barotropic model of §11.1 still involves a number of approximations. In particular the motion is assumed to be quasi-geostrophic. This restriction can be removed by writing the equations in a more basic form, namely: the horizontal momentum equations (7.8)

$$\frac{\partial \mathbf{V}}{\partial t} + \mathbf{V} \cdot \mathbf{\nabla}V + \omega\frac{\partial \mathbf{V}}{\partial p} + f\mathbf{k} \wedge \mathbf{V} = -\mathbf{\nabla}\Phi + \mathbf{F} \qquad (11.12)$$

the continuity equation (7.20)

$$\mathbf{\nabla} \cdot \mathbf{V} + \frac{\partial \omega}{\partial p} = 0 \qquad (11.13)$$

the hydrostatic equation (1.2)

$$\frac{\partial \Phi}{\partial p} + \frac{1}{\rho} = 0 \qquad (11.14)$$

and the thermodynamic equation (11.6)

$$\frac{\partial}{\partial t}\left(-\frac{\partial \Phi}{\partial p}\right) + \mathbf{V} \cdot \mathbf{\nabla}\left(-\frac{\partial \Phi}{\partial p}\right) - \frac{B\omega}{g\rho^2} = \frac{1}{c_p\rho}\frac{dS}{dt} \qquad (11.15)$$

In these equations $\mathbf{V}$ is the horizontal wind and ∇ refers to differentiation at constant pressure.

The five equations (11.12) to (11.15) ((11.12) is of course two equations) involve five unknowns, two components of $\mathbf{V}$, ω, Φ and ρ.

Although ordinary sound waves have been eliminated from the solutions of this set of equations by the use of the hydrostatic equation (11.14) and the neglect of vertical accelerations, solutions remain for a wide range of different motions including for instance the acoustic and the gravity waves described in §8.3. These solutions are eliminated for the simplified models of §11.1 and §11.2 through the quasi-geostrophic approximation.

With primitive equation models care has to be taken that the solutions are not swamped by spurious gravity or acoustic waves which may arise from errors in the initial data or from computational instability. To overcome these problems it is necessary to introduce an initial motion field which satisfies (11.12); actual observations of wind must not be employed as they stand. Further a time step has to be chosen for the integration process which is short enough to avoid computational instability for the fast-moving gravity or acoustic waves. For this reason time steps need to be considerably shorter than with quasi-geostrophic models with similar horizontal resolution. Primitive equation models, therefore, demand much more computer time than comparable quasi-geostrophic ones.

The scheme for integration of the dynamical equations needs to be designed to conserve mass, mass-weighted potential temperature and moisture, and angular momentum. Especially are these conservation requirements important when long integrations are carried out, for instance for investigating changes in climate (cf. chapter 14).

11.4 Parametrizations

In (11.12) a frictional term **F** has been included and in (11.15) the diabatic heating term. Many of the physical processes involved in these terms have been introduced in earlier chapters. These physical processes that occur in the free atmosphere or close to the surface are generally complex and involve space scales that are much less than a typical grid size of the model. The descriptions of these processes that can be fitted into the scope of a model and the available computing capacity can only involve a few parameters and relatively simple algorithms. Such descriptions are called *parametrizations*. The challenge is to capture the essential physics especially as it affects the large scale flow. Before providing, in later sections of this chapter, outlines of the most important parametrizations, some brief details will be given of the construction of modern forecasting models.

11.5 Models for weather forecasting

As an example of the models currently employed at a major forecasting centre, the Unified Model for the atmosphere in use at the UK Meteorological Office, Bracknell, will be briefly described. It is called 'unified' because it employs essentially the same formulation for three different purposes. The Global Forecast Model possesses a horizontal

Fig. 11.2. The horizontal grids used by the Global and Mesoscale Models of the UK Meteorological Office. For the global grid every second point is shown and for the UK grid every fourth point.

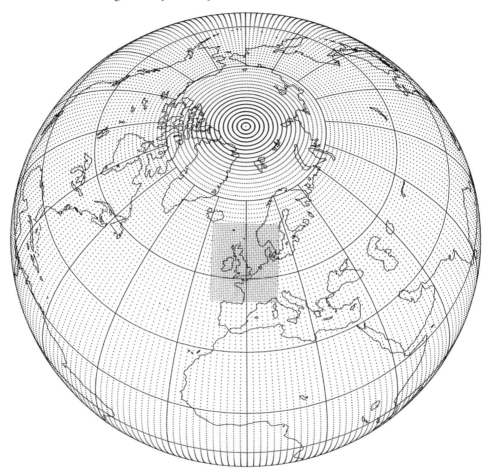

resolution of 0.83° (longitude) × 0.56° (latitude), the Mesoscale Model covers a limited region with a horizontal resolution of 12 km (fig. 11.2) and the atmospheric component of the model employed for climate prediction (cf. section 14.7) possesses a horizontal resolution of 3.75° × 2.5°. There are 30 or 38 levels in the vertical grid which is terrain following near the surface (§11.6) but evolves to constant pressure surfaces at higher levels. Because of the importance of the boundary layer, a number of levels close together are included near the surface. The time steps for the integrations in the three models are 20, 5 and 30 minutes respectively for computations of the physics; sweeps of the dynamical computations are between 2 and

Fig. 11.3. Schematic diagram illustrating the data assimilation process. The ordinate ψ represents the state of the model or the atmosphere. T is the synoptic time appropriate to the observations to be assimilated. $T-6$ is the start of the assimilation process from the model prediction for the time $T-6$ based on the previous forecast run. $\psi^{observed}$ is the state of the atmosphere as determined from observations at time T; ψ^{model} is the model prediction for time T based on the previous forecast run. The line ending $\psi^{model+data}$ illustrates the assimilation process. At each time step (steps are separated by Δt) the model is modified to move in small increments towards the observations. The assimilation process for asynoptic observations may be applied at the correct time for each observation. (After Atkins & Woodage, 1985)

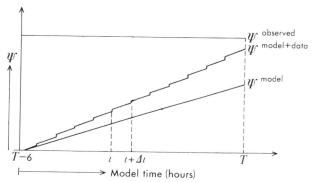

4 times more frequent. At high latitudes where the grid spacing in longitude becomes small, a filtering process is introduced to ensure computational stability. Boundary conditions for the mesoscale forecasts are provided from global forecast fields interpolated to the Mesoscale Model co-ordinates. Details of the dynamical scheme for the Unified Model are given by Cullen and Davies (1991) and White and Bromley (1995).

Observations cannot be inserted into the model without considerable care. The process of *data assimilation* inserts data gradually (fig. 11.3) using an *Analysis Correction Scheme*. This is an iterative analysis process whose purpose is to maintain dynamical balance of the model as the data assimilation proceeds (for more details see Lorenc et al., 1991).

Despite the increase in the speed of computers (fig. 11.4), it remains the case that large forecasting models are computer limited. Faster computers, for instance, would allow for higher horizontal and vertical resolution in the models. A 24-hour forecast with the global model involves about 5×10^{12} numerical operations and takes 7 minutes using 144 processors on the CRAY T3E-9OO computer. The computer produces forecast output for many different applications. An example of Global Model output is shown in fig. 11.5 and of a particular model forecast in fig. 11.6.

Fig. 11.4. The growth in speed in number of floating point operations per second of computers employed by the UK Meteorological Office for numerical forecasting. A straight line has also been drawn with a growth of a factor of 10 every 4.5 years. Richardson's 'computer' refers to his dream (Richardson, 1922) of a building filled with a large number of people organized so as to carry out the computations for weather forecasting.

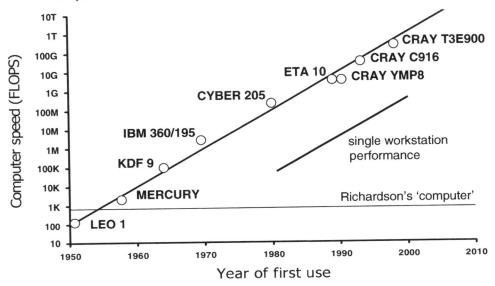

11.6 Inclusion of orography

It is necessary in a model to take into account the varying height of the surface. In the isobaric co-ordinate system where pressure is the vertical co-ordinate, the varying surface pressure due to the varying height of the surface is not easy to handle. To overcome this problem, numerical models employ a co-ordinate σ defined as

$$\sigma(x, y, t) = \frac{p(x, y, t)}{p_s(x, y, t)} \qquad (11.16)$$

where p_s is the surface pressure. This system is known as the sigma co-ordinate system (fig. 11.7, problem 11.1). σ is a non-dimensional parameter; the lower boundary is at $\sigma = 1$ where the vertical velocity $\dot{\sigma} = 0$.

11.7 Radiation transfer

Radiative processes are examples of the physical processes that contribute to the diabatic term in (11.15) and for which appropriate parametrization is required. They divide into two parts (§2.1) namely the absorption and scattering of solar radiation and the exchange of terrestrial radiation. Interacting with the radiation streams are molecules, aerosol

Fig. 11.5. A chart showing upper winds (at 39000ft) and temperatures issued for the use of aviation directly from the global numerical forecast model at the UK Meteorological Office. Temperatures are in −°C: the wind at 30°N, 40°W is 65 kt from the NW.

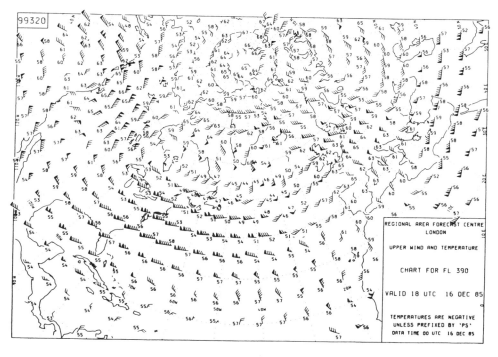

particles and clouds which may be present in multilayers. The basic equations governing radiative transfer for both solar and terrestrial radiation were formulated in chapters 4 and 6 where simple examples of radiative transfer calculations were also given.

The detailed structure of the molecular bands of the gases present in the atmosphere are well known and well documented. Some summary information can be found in appendices 8, 9 and 10. The routine computations in numerical models are simplified in three ways, namely:

(1) both scattered solar radiation and terrestrial radiation are divided into two streams, upwards and downwards (cf. §§2.2 & 6.4) with appropriate scaling of the paths of absorbing and emitting gases;

(2) empirical transmission functions (cf. fig. 11.8), normally expressed as a sum of exponentials, are developed for different molecular bands (or parts of bands) appropriate to atmospheric paths, by comparing against detailed line by line integrations over frequency;

Fig. 11.6. Illustrating a 24-hour forecast with the global model of the UK Meteorological Office for the 'Boxing Day' storm of 1998 that brought damage and disruption to parts of the UK: (a) surface pressure for midday on 25 December 1998 illustrating that there was no pre-existing storm; (b) surface pressure analysis for midday on 26 December 1998; (c) surface pressure for midday on 26 December 1998 as forecast from (a). (Provided by A. Lorenc & F. Saunders, UK Meteorological Office.)

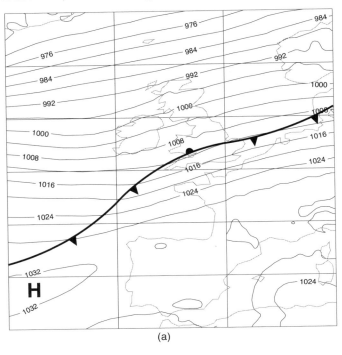

(a)

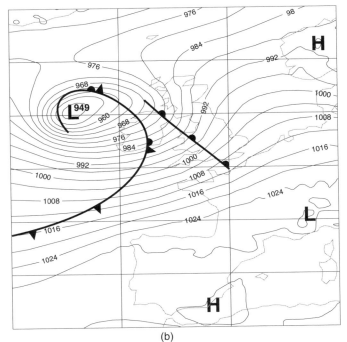

(b)

Fig. 11.6. (*continued*)

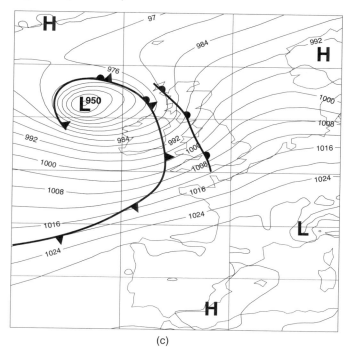

(c)

Fig. 11.7. The σ–co-ordinate system for a 15-level Global Circulation Model.

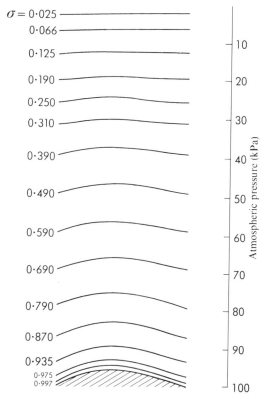

Fig. 11.8. Illustrating that pressure scaling is a good approximation. (a) The absorption of solar radiation by an atmosphere containing water vapour as a function of modified path length u* given by

$$u^* = \sec\theta \int_{\rho\&t} c\rho(p/p_0)^x \, dz$$

for a range of values of p/p_0 and where $x = 0.675$ is chosen to give the best fit. (After Manabe & Wetherald, 1967)

(b) The same as (a) but for an atmosphere containing carbon dioxide with modified path length u^* (in $g\,cm^{-2}$) and with $x = 0.68$.

(c) The emissivity of a slab of atmosphere at 220 K containing water vapour showing the usefulness of the modified path length u^* ($x = 0.7$) in $g\,cm^{-2}$ as a parameter. The wavenumber range 550–800 cm^{-1} has been omitted since in that range CO_2 absorption is dominant. (All after Manabe & Wetherald, 1967).

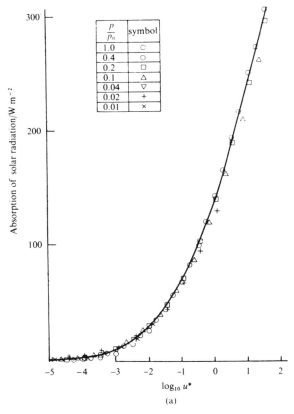

(a)

(3)   empirical scaling approximations are also developed (again by comparing with exact line by line calculations as in (2) above) to be applied to absorber amounts to allow for the pressure and temperature dependencies of absorption (fig. 11.8).

Further details of these simplifications can be found in Goody and Yung (1989), and Edwards and Slingo (1996). The errors introduced by these

Fig. 11.8. (*continued*)

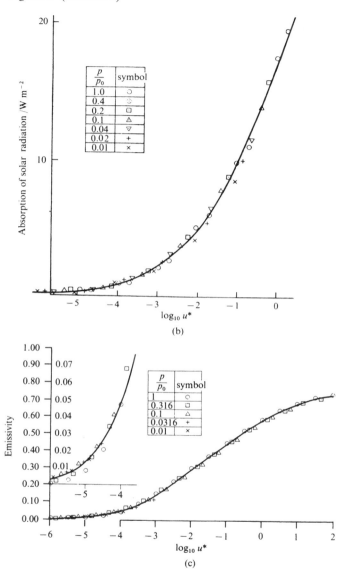

(b)

(c)

approximations are generally much smaller than those which arise from errors in data or in the specification of other atmospheric parameters, for instance those involved with the description of clouds.

11.8   Sub grid scale processes

We saw in chapter 9 that interaction in the atmosphere occurs between motions on a wide range of scales. With a numerical model, therefore, that inevitably possesses a finite grid size it is necessary to make allowance for the effect of motions on scales smaller than the grid size. The term **F** in (11.12) was included to allow for the transfer of momentum by small scale eddies; a similar term needs to be included to allow for the eddy transfer of water vapour. Further, there is a contribution to the term dS/dt in (11.15) from the eddy transfer of heat. The simplest formulation of both horizontal and vertical eddy transfer employs eddy diffusion coefficients K which when multiplied by the appropriate gradients give the transfer, e.g. (9.6), (9.23) and (9.24). Typical values of K for the troposphere for sub grid scale parametrization lie in the range $1-10\,m^2 s^{-1}$. The Ks are allowed to vary with height, the static stability, or with quantities describing the local flow. In particular, K for momentum can be formulated in a way that preserves the k^{-3} spectral distribution of turbulent energy (§9.6). More detail of Ks that vary with static stability is given in §11.9.

11.9   Moist processes and clouds

Water vapour evaporated from the surface and distributed through-out the atmosphere by atmospheric motions plays a very significant part in the atmosphere's energy budget because of (1) the release of latent heat of condensation and (2) the formation of clouds which alter the radiation budget (cf. chapter 6).

Early models dealt with clouds very crudely. Typically, model clouds were generated at specified levels whenever the relative humidity exceeded a critical value, chosen for broad agreement between model gen-erated cloud cover and that observed from climatological records. The type and amount of cloud was specified empirically, again using climatological data. Immediate precipitation was assumed of the water condensed by returning the humidity to its critical value.

More recent models parametrize the processes of condensation, freezing, precipitation and cloud formation much more completely. They also provide model descriptions of cloud properties that enable their radia-tive properties to be specified sufficiently well for the influence of clouds on the atmosphere's overall energy budget to be properly included.

Layer clouds are dealt with separately from convective clouds. The latter, which in general are of smaller scale than a grid box, are considered in the next section.

For the parametrization of clouds, their formation and their prop-erties, a key prognostic variable q_c representing cloud water content needs to be added to the model scheme. q_c can be either in the liquid (q_L) or ice (q_F) phase or in a mixture of both depending on the cloud temperature. A thermodynamic variable T_L and a water content variable q_w that are conserved during changes of state of cloud water can be defined as follows:

$$T_L = T - (L_C/c_p)q_L - \{(L_C + L_F)/c_p\}q_F \qquad (11.17)$$

$$q_W = q + q_C = q + q_L + q_F \qquad (11.18)$$

where L_C and L_F are respectively the latent heats of condensation and fusion, q_w is the total water content and q the water vapour content. Vertical turbulent mixing of these conserved quantities, dependent on the local Richardson number (see problem 9.7), is assumed to occur in the cloudy region.

From these variables, the fraction of cloud cover and the cloud water content required is generated. The scheme described by Smith (1990) is based on assumptions about the sub grid scale probability distribution of the conserved variables defined in (11.17) and (11.18) and about the critical relative humidity for cloud formation (typical values assumed fall in the range 0.85–0.95).

Parametrization is also required of the rate of conversion of cloud water into precipitation which is mainly dependent on the cloud water content. It is also different for water and ice clouds. For ice clouds, for instance, the particles are larger and are assumed to possess a constant fall speed. Evaporation of precipitation at levels below the cloud can also be allowed for.

Finally, to describe adequately the influence of clouds on the transfer of radiation in the atmosphere, scattering, absorption and emission properties of the clouds need to be included in the calculations described in §11.7. For this to be done, the mean number density and size of water drops or ice crystals in the clouds must be specified; these will be dependent on the liquid water or ice content and fractional cloud cover. The shape of the ice crystals is also needed; the simplest assumption is that they are spherical, although, for cirrus clouds, allowance can be made for more elaborate shapes. A comparison of the cloud radiation forcing deduced from model simulations of clouds using different parametrization schemes and the observed radiative forcing is shown in fig. 6.5.

11.10 Convection

Cumulus convection plays an important role in the maintenance of the atmosphere's large scale circulation. Early models introduced convection very simply through a process of *convective adjustment*. The temperature profile at any location in the model is adjusted so that it does not exceed the dry adiabatic (§1.4) under conditions of low water vapour content and so that it equals the saturated adiabatic (§3.2) when the relative humidity is equal to or close to 100%. This adjustment has to be made, of course, in such a way that the total energy is unchanged, i.e. the sum of the total potential energy and latent heat released by the adjustment must be zero.

This simple convective adjustment process, however, does not provide an adequate estimate of the net energy release or a sufficiently realistic estimate of the distribution of the released energy in the vertical and hence of its influence on the larger scale circulation. A parametrization scheme is required which represents more of the essential details of the convective process. Single convective cells, however, cannot be represented in numerical models because they are much smaller than the model grid. It is therefore necessary to treat an ensemble of convective clouds and to represent their net effect on the atmosphere in terms of grid scale parameters.

The convective parametrization scheme is based on a single air parcel representing an average over the ensemble. For a column of atmosphere, working from the bottom upwards, each layer of the model is tested until one is found with excess buoyancy. The convective process (fig. 11.9, and Gregory & Rowntree, 1990) is then initiated, a parcel of air continuing to rise through the layers in the model, entraining air from the envi-

Fig. 11.9. Schematic of the parametrization of convection in an atmospheric model. (From the Unified Model of the UK Meteorological Office).

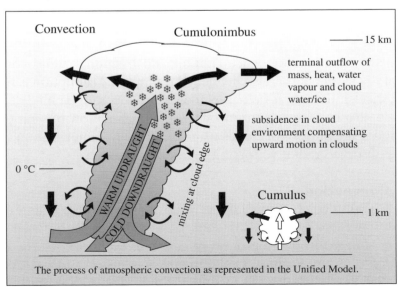

The process of atmospheric convection as represented in the Unified Model.

ronment and detraining cloudy air until it is no longer buoyant. The upward mass flux is dependent on the excess buoyancy and is modified as the parcel rises by the entrainment and detrainment processes. The cloud properties within the rising parcel are determined through parametrization of the processes of condensation, mixing and precipitation in similar ways to those described in §11.9. Specific treatment of downdrafts adds to the realism of the scheme.

11.11 Transfer across the surface

Momentum, heat and water vapour are transferred across the earth's surface; parametrizations for the transfer of each are required.

First, a method is required to deduce the surface temperature. Over the ocean, for weather forecasting models, since it changes only slowly the surface temperature is specified from observations. For seasonal forecasting or climate models that couple the atmosphere to the ocean circulation (or to part of it) the ocean temperature is generated by the coupled model (cf. fig. 14.11). Over the land surface, for which the effective heat capacity is small, the surface temperature T_s is determined by the local heat balance, i.e.

$$(1-A)I_s + F^\downarrow - \sigma T_s^4 + Q_s - LE = 0 \qquad (11.19)$$

where σ is the Stefan–Boltzmann constant, A the surface albedo, I_s the incident solar radiation (after allowing for absorption or scattering in the atmosphere or in the clouds above as described in §11.7), $F^\downarrow$ the downward flux of terrestrial radiation from the atmosphere or from clouds, again as described in §11.7, Q_s the vertical transfer of heat from the atmosphere to the surface by eddy processes, E the rate of evaporation of water vapour from the surface, and L the appropriate latent heat. The emissivity of the surface for terrestrial radiation is assumed to be unity. Methods of calculating Q_s and E are described below.

If the model's lowest level is close to the surface and well within the boundary layer, the simplest expression for the flux of momentum across the surface or the surface stress τ is

$$\tau = -\rho c_D |V|\, V \qquad (11.20)$$

where ρ is the density, c_D the drag coefficient (see problem 11.3), and V the vector wind appropriate to the first level. Similar expressions may be written for the flux of heat Q_s and water vapour E, namely

$$Q_s = c_p \rho c_D |V| \Delta\theta \qquad (11.21)$$

and

$$E = \rho c_{\mathrm{D}} |V| \Delta m \qquad (11.22)$$

where $\Delta\theta$ and Δm are respectively the difference in potential temperature and water vapour mixing ratio between the surface (assumed saturated at surface temperature) and the first level.

Equations (11.20) to (11.22) are applicable to conditions of neutral stability; the effect of different stabilities can be taken into account by allowing c_{D} to vary with stability. Equation (11.22) also only applies to a wet surface. This is, of course, all right over the ocean. Over the land, allowance must be made for variation in the soil moisture content W (mass per unit area) which can be found from the equation

$$\frac{\partial W}{\partial t} = P - E \qquad (11.23)$$

where P is the precipitation rate determined from the moist processes of §11.9. If W exceeds a value specified for a given location the excess water is assumed to run off. If W exceeds a certain lower value W_k the surface is assumed to be wet; if $W < W_k$ the evaporation rate is assumed to be reduced by the ratio W/W_k from the value given by (11.22). Snow cover can be allowed for in a similar way to soil moisture.

The simple formulation of equation (11.20) applies reasonably well over the sea or level terrain. To estimate the momentum exchange due to flow over hills or mountains, it is necessary to model the generation of gravity waves (§8.3 and problem 11.5), their propagation upwards depending on the vertical stability and the subsequent absorption of their momentum in the upper troposphere and lower stratosphere where the waves break (Palmer et al., 1986, and Milton & Wilson, 1996).

11.12 Forecast model skill

The skill of a forecast model can be measured by comparing the forecast meteorological synoptic analysis for a given forecast period with that derived from observations. In fig. 11.10 is shown, for the past 30 years, the improvement in skill for different forecast periods for the forecasting model of the UK Meteorological Office. It demonstrates that gradual improvements in the representation of the model dynamics and physics and improvements in the model resolution have led to steady improvements in skill. In chapter 13 (§13.3) the question will be addressed as to how far such improvements can be expected to continue.

11.13 Other models

The models so far described in this chapter are *grid-point models*, in which model parameters are specified at grid points and integrations

Fig. 11.10. Errors (root mean square differences of forecasts of surface pressure compared with analyses from observations) of UK Meteorological Office forecasting models for the region of the North Atlantic and Western Europe from 1967 to 1998 for 24-, 48- and 72-hour forecasts (from Meteorological Office).

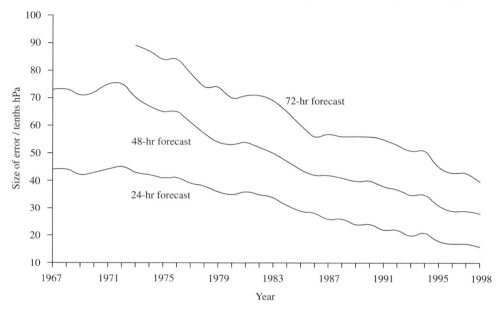

carried out by finite difference methods applied to the array of grid points. Another type of model which possesses some advantages and which is widely employed is the *spectral model* in which horizontal variations are represented in terms of series of spherical harmonics. Such models are particularly valuable for studies of the upper atmosphere. Here most of the variations are described by the first few terms of such a series, i.e. those describing the first few wavenumbers (the wavenumber refers to the number of complete cycles of a variation which occur around a latitude circle). A substantial amount of computing time can therefore be saved by the use of such transformations.

Also valuable for particular studies are *two dimensional models*, i.e. models in which atmospheric parameters are averaged around latitude circles so that only variations with height and latitude are considered. The main difficulty with such models is that of devising means for dealing with the large scale eddy transports of heat and momentum which have to be related in simple ways to zonally averaged quantities. Two dimensional models are particularly useful in upper atmospheric modelling in which changes due to complex photochemical reactions (cf. §5.6) have to be included and for which integrations over long periods (e.g. several years) are required (see e.g. Harwood & Pyle, 1975).

The models we have described in this chapter have shown a great deal of success in describing the dynamics of the atmosphere and in operational forecasting. Nevertheless, these models contain some fundamentally unsatisfactory features. For instance in §11.3 it is pointed out that the numerical solutions derived from the set of primitive equations not only describe the development of the relatively slowly varying *synoptic* field which is the primary interest of the weather forecaster, but they also describe a much wider class of solutions including gravity waves. Means of damping out these unwanted solutions have to be applied, a process which becomes increasingly difficult as the horizontal resolution of the model is increased. This higher resolution is needed as was pointed out in §11.5 in order to include explicitly as much as possible of the interaction between the synoptic flow and the smaller scale detail and to attempt to describe more adequately the almost discontinuous structures (e.g. fronts, inversions and the tropopause) which are, in effect, part of the synoptic pattern. However, the highest resolutions currently realized in models do not match the scale of these structures.

Current models distinguish between resolved and unresolved features purely on the basis of horizontal scale and time scale. A procedure which would overcome this difficulty would be to employ a *Lagrangian*[†] description of the air parcels allowing their properties to evolve slowly, but without restricting the spatial scale. It has been shown by Cullen & Purser (1984) that an effective means of describing the evolution is to employ an energy principle which assumes that the air parcels are arranged at all times close to a minimum energy configuration. The configuration evolves slowly in time under the action of the forcing fields. The method has so far been successfully applied to the modelling of fronts. Further details are given in problems 11.7 and 11.8.

Problems

11.1 Show that the following equation will transform from the isobaric vertical co-ordinate system to the sigma system

$$\mathbf{\nabla}_p = \mathbf{\nabla}_\sigma - \frac{\sigma}{p_s}\,\mathbf{\nabla}p_s\,\frac{\partial}{\partial\sigma} \tag{11.24}$$

[†] A *Lagrangian* description is one which follows the flow of the fluid as opposed to an *Eulerian* description which describes the flow referred to a co-ordinate system fixed in space.

where ∇ refers to horizontal differentiation only. Hence derive the horizontal momentum equation

$$\frac{dV}{dt} + f\mathbf{k} \wedge \mathbf{V} = -\nabla\Phi + \frac{\sigma}{p_s}\nabla p_s\frac{\partial\Phi}{\partial\sigma} \qquad (11.25)$$

and the continuity equation

$$\nabla\cdot(p_s\mathbf{V}) + p_s\frac{\partial\dot\sigma}{\partial\sigma} + \frac{\partial p_s}{\partial t} = 0 \qquad (11.26)$$

where $\dot\sigma = \dfrac{d\sigma}{dt}$

11.2 Consider a vertical column of unit cross-section above a location where the surface pressure is p_s. By considering the convergence of mass into the column show that

$$\frac{\partial p_s}{\partial t} = -\int_0^1\nabla\cdot(p_s\mathbf{V})d\sigma$$

Show that the same result may be obtained by integrating the continuity equation (11.26).

11.3 Show from (9.15) that under neutral conditions c_D in (11.20) is given by

$$c_D = \left[\frac{\kappa}{\ln(z/z_0)}\right]^2 \qquad (11.27)$$

Hence find its value if $z = 100\,\text{m}$, $z_0 = 1\,\text{cm}$ and $\kappa = 0.4$.

11.4 A uniform cloud of reflectivity R is present above a surface with albedo α. Taking into account multiple reflections between the cloud and the surface and ignoring absorption or variations of reflectivity with angle show that if the proportion of the incident solar flux S absorbed by the surface is written as $(1 - \alpha_e)S$ where α_e is an effective reflectivity, then

$$\alpha_e = \frac{\alpha(1-R)}{(1-\alpha R)} \qquad (11.28)$$

11.5 The generation of gravity waves by mountains is an important mechanism leading to exchange of momentum between the atmosphere and the surface. If waves are generated with horizontal and vertical velocities u and w (both varying as $\exp i(\omega t + kx + mz)$) the stress τ_s at the surface will be given by equation (9.6)

$$\tau_s = \overline{\rho u w}$$

Show that the vertical velocity w generated by a mountain of height h will be proportional to $k\bar{u}h$. Then employing equation (8.24) and the result of problem 8.21, show that providing $m \gg \frac{1}{2}H$ and $m \gg k$

$$\tau_s = \alpha k \rho \bar{u} N_B \bar{h}^2 \qquad (11.29)$$

where α is a constant which may be chosen empirically.

In the formulation of gravity wave drag for incorporation into the Meteorological Office numerical model Palmer et al. (1986) have given αk the value $2.5 \times 10^{-5}\,\text{m}^{-1}$ and limited the value of $\bar{h}^2$ to $(400\,\text{m})^2$. Consider some real orography and discuss the reasonableness of these values.

11.6 A height-like variable z can be defined such that it is related to the pressure p by the expression

$$z = \frac{H_0}{\kappa}\left(1 - \left(\frac{p}{p_0}\right)^{\kappa}\right) \qquad (11.30)$$

where $H_0 = RT_0/Mg$ is the scale height under surface conditions where $p = p_0$ and the temperature $T = T_0$ and where $\kappa = (\gamma - 1)/\gamma$ (cf. §3.1).

Show from the hydrostatic equation and equation (3.4) that increments in z are related to increments in geometrical height h by the expression

$$\theta dz = \theta_0 dh \qquad (11.31)$$

where θ is the potential temperature and $\theta_0 (= T_0)$ is the potential temperature at the surface. Hence show that for an adiabatic atmosphere $z = h$. Show also that the maximum value of z is about 28 km. What is the percentage difference between z and h for an isothermal atmosphere at the level where $z = H_0$?

11.7 For an incompressible, hydrostatic, atmosphere under the Boussinesq assumption (i.e. the atmosphere is assumed to be of uniform density apart from buoyancy effects) and with the vertical co-ordinate z defined as in problem 11.6, derive the following set of equations.

$$\frac{dM}{dt} = -\frac{\partial \phi}{\partial y} \qquad (11.32)$$

$$\frac{dN}{dt} = \frac{\partial \phi}{\partial x} \qquad (11.33)$$

$$\frac{\partial \phi}{dz} = \frac{g\theta}{\theta_0} \qquad (11.34)$$

$$\frac{d\theta}{dt} = 0 \qquad (11.35)$$

$$\frac{\partial u}{\partial x} + \frac{\partial v}{\partial y} + \frac{\partial w}{\partial z} = 0 \qquad (11.36)$$

Here ϕ is the geopotential, θ the potential temperature and

$$M = v + fx \qquad (11.37)$$

and $\quad N = -u + fy$

are components of what is called the *absolute momentum* (Hoskins, 1975). They are given that name because, taking into account the rotation of the earth, in the absence of forcing terms (e.g. pressure gradients), M and N are the appropriate conserved quantities which replace the velocities v and u. Appropriate boundary conditions to go with the set of equations (11.32)–(11.36) are $w = 0$ at $z = 0$ and H.

In deriving an expression for the energy, it is useful to remember that for a fluid of nearly uniform density given by $\rho(h)$ at level h, the difference between its potential energy and that which would occur if the fluid were of completely uniform density is given by

$$\int \frac{g(\rho(z) - \rho_0)h}{\rho_0} \rho_0 dV$$

Hence show that for a Boussinesq atmosphere under the assumptions being made here, an appropriate expression for the sum E of the kinetic and potential energies is:

$$E = \int_V \left[\frac{1}{2}(u^2 + v^2) - \frac{g\theta z}{\theta_0} \right] \rho_0 dV \qquad (11.38)$$

where the integral is over the whole domain of interest. By substituting for M and N show that E can be written

$$E = \int \left[\frac{1}{2}f^2(y^2 + x^2) + \frac{1}{2}(M^2 + N^2) - Nfy - Mfx - \frac{g\theta z}{\theta_0} \right] \rho_0 dV \qquad (11.39)$$

The procedure which is being developed by Cullen et al. (1986) is to determine the evolution of the flow from a balanced initial state by minimizing the energy integral (11.39) in a way which is consistent with the equations (11.32)–(11.36). The volume of an air parcel in x, y, z space is conserved by (11.36), the potential temperature

θ of the parcel is conserved by (11.35). No fluid is allowed to cross the boundaries. The evolution of the flow is driven by the pressure gradient terms in (11.32) and (11.33) and the fluid responds by rearranging itself in such a way as to minimize the energy.

The simplest case to consider is one dimensional in which there is no horizontal motion and $\theta = \theta(z)$ only. The energy is then potential energy only. Show by exchanging an air parcel near the surface with one at a higher level that the potential energy term in (11.39) is minimized by moving to an arrangement in which θ increases monotonically with z. This is similar to the process described in §3.5 when considering available potential energy.

Now going back to equation (11.39) we note that the first term in the energy integral is independent of the flow and the second does not change under rearrangement because, as we mentioned above, in the absence of forcing M and N are conserved. It is therefore required to minimize the remaining terms, i.e.

$$\int \left(-Nfy - Mfx - \frac{g\theta z}{\theta_0} \right) \rho_0 dV \tag{11.40}$$

By extension of the one dimensional argument, it can be shown that for (11.40) to be minimized under all possible rearrangements it is necessary for N to increase monotonically with y, M with x and θ with z.

An important theorem proved by Cullen and Purser (1984) is that there exists a unique minimizing arrangement for a fluid divided up into finite elements in which the quantities M, N and θ in (11.40) must be related to a parameter η such that

$$fM = \frac{\partial \eta}{\partial x} \quad fN = \frac{\partial \eta}{\partial y}, \quad \frac{g\theta}{\theta_0} = \frac{\partial \eta}{\partial z} \tag{11.41}$$

Show that if

$$\eta = \phi + \frac{1}{2} f^2 \left(x^2 + y^2 \right) \tag{11.42}$$

then M and N are as defined in equation (11.37) but with u and v replaced by their geostrophic values.

With M and N defined in this way, in an arrangement with finite elements in each of which the values of M, N and θ are constant, show that the slopes of the boundary between two elements whose values of M, N and θ differ by ΔM, ΔN and $\Delta \theta$ respectively are given by

Fig. 11.11. Illustrating a very simple model of a sea breeze front.

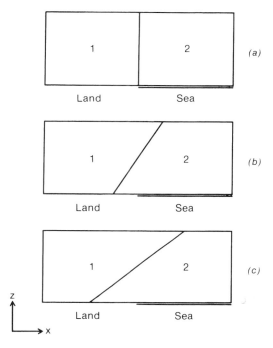

$$\frac{dx}{dz} = \frac{g}{f\theta_0} \frac{\Delta\theta}{\Delta M} \tag{11.43}$$

$$\frac{dy}{dz} = \frac{g}{f\theta_0} \frac{\Delta\theta}{\Delta N} \tag{11.44}$$

Note that equations (11.43) and (11.44) are essentially the thermal wind equations (cf. §7.6).

11.8 To illustrate the application of the methods described in problem 11.7, consider an atmosphere divided into two finite elements of equal volume (fig. 11.11) initially at rest, possessing identical values of θ and possessing values of M differing only because the mean value of x differs between the two elements. Initially, the lower boundary of one element is land, the other sea. As the land heats up during the day, show that the boundary between the two elements (i.e. the sea breeze front) changes in the way shown in fig. 11.11 until an equilibrium state is reached (fig. 11.11(c)) in which the two elements contain equal areas of the land surface.

12

Global observation

12.1 What observations are required?

To provide adequate and accurate initial conditions for numerical models and to enable comparison to be made between models and the real atmosphere, detailed knowledge of the state of the whole atmosphere is required continuously in time. We need to know the three dimensional fields of motion, density and composition together with the appropriate conditions at the surface. Clearly measurement in complete detail is not possible. Further, some parameters are related to each other. For instance, away from the equatorial regions and the boundary layer, the geostrophic approximation is a good one so that motion need not necessarily be measured independently if the pressure field is specified and vice-versa. Also, if the field of temperature is known together with the pressure at some reference surface, the density and pressure can be deduced at all levels from the hydrostatic equation (1.4) (problem 12.1).

The following is a typical specification of the observations required for global studies; it is a shortened version of that prepared in 1979 for the first Global Atmospheric Research Programme global experiment.

Atmospheric state parameters	Accuracy (RMS error)
Wind components	$\pm 3 \, \mathrm{m \, s^{-1}}$
Temperature	$\pm 1 \, \mathrm{K}$
Pressure of reference level	$\pm 0.3 \, \%$
Water vapour pressure	$\pm 0.1 \, \mathrm{kPa}$
Sea surface temperature	$\pm 0.2 \, \mathrm{K}$

Measurements of these parameters (or their deduction from other observations) are required every 12 hours, with at least one measurement every 100 km in the horizontal and at least eight data levels in the vertical (surface, 90, 70, 50, 20, 10, 5 and 2 kPa). Information is also required on precipitation, cloud cover, surface conditions (e.g. snow or ice cover, albedo) and elements of the radiation budget.

For studies on a local scale rather than the global, and for detailed studies of particular processes, measurements with more intense coverage both in space and time and with more accuracy are required. The value of improvements in data is explained in §13.3. For studies on very long time scales, for instance of climate change, a more comprehensive set of measurements is needed including parameters describing the ocean circulation (see §14.8).

12.2 In situ observations

Over the land areas of the world much of the information about the atmosphere near the surface comes from a high density of conventional observations of pressure, temperature, humidity, wind, cloud cover, and precipitation made under standard conditions of exposure. Over the oceans, similar observations near the surface, including also the important parameter of sea surface temperature, are made from ships and from fixed and free-floating buoys.

Above the surface measurements are made twice per day from a large number of *radiosondes*. These are balloon-borne packages containing simple instruments (fig. 12.1) for the measurement of pressure, temperature and humidity, together with a radio transmitter and a radar reflector so that the package can be tracked to give details of the wind. Measurements up to ~30 km (~1 kPa) are possible with the radiosonde. A much sparser network of rocket-sondes – sondes released from high altitude (up to ~90 km) rockets which then descend on parachutes – has enabled some observations to be made at higher levels.

More specialized measurements, especially of atmospheric composition (e.g. the concentration of ozone or other minor constituents), are possible from larger balloons, aircraft or rockets.

12.3 Remote sensing

The technique of *remote sensing* (also called *remote sounding*) employs electromagnetic radiation reflected, scattered or emitted from the atmosphere at a distance from the observing station to infer atmospheric structure or composition. *Active* remote sensing employs a transmitter at the observing site, while *passive* remote sensing relies on the radiation emitted from the atmosphere or on solar radiation scattered or reflected by it towards the observer.

Fig. 12.1. Compatibility between observations from different meteorological instruments is of paramount importance as observations from many different sources provide the input for the global analyses from which global forecasting models begin their integrations. Illustrated here are five radiosondes (which measure pressure, temperature, humidity, wind) used by different countries attached to the same balloon so that direct comparison between their measurements can be made. (Courtesy of Meteorological Office)

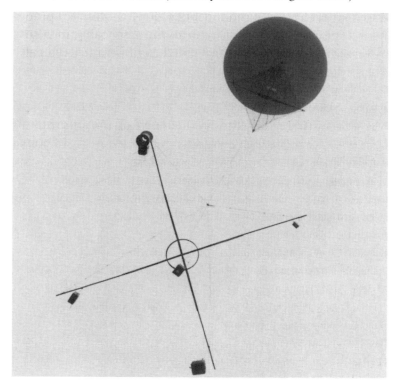

Some of the earliest classic examples of remote sounding are those of the structure and composition of the high atmosphere made by observers on the surface. It was observations of meteor trails by Lindemann and Dobson in 1923 which first demonstrated the high temperature near the stratopause and observations of the spectrum of ultraviolet light scattered from the sky by Dobson in 1926 which provided the first measurements of stratospheric ozone and its distribution with height (see §12.8 for a description of this technique as used in observations from satellites).

12.4 Radar and lidar observations from the surface

Radar, typically around 3 cm in wavelength, is an important active remote sensing technique for the observation of precipitation and its distribution (fig. 12.2). Water drops and ice crystals scatter back radar energy, the scattered intensity being strongly dependent on droplet size (to the sixth power of the diameter) and on whether the precipitation particles are water or ice. Multi-frequency observations can assist in sorting out some of these dependencies. Calibration factors, preferably measured directly by comparing some radar signals with actual rainfall rates, need to be applied to convert a map of radar echoes to one of rates of precipitation.

Radar, directed upwards close to the vertical, can also provide detailed information about the wind field. Wavelengths typically in the range 30 cm to a few metres are employed, the backscatter coming from small scale inhomogeneities in refractive index due to variations in water vapour or temperature produced by turbulence or atmospheric waves. *Wind profilers* are small radar installations making routine wind observations up to about 8 km altitude. *MST (mesosphere, stratosphere and troposphere) radars* are more powerful installations for particular studies of wind structure (fig. 12.3) and turbulence at all levels up to the mesosphere; above 60 km in the daytime the major contribution to backscatter is from the electron concentration.

Lidars provide intense sources of monochromatic radiation which can be directed upwards from the surface. The analysis of backscattered radiation from them can provide information about the distribution of clouds, aerosol and some minor constituents.

12.5 Remote sounding from satellites

The great advantage of a satellite as a measurement platform is that it provides for coverage in space and time on a scale that is not otherwise achievable. From a satellite in *geostationary* orbit at 35 000 km altitude directly above a fixed point on the equator, continuous observation of about a quarter of the atmosphere is possible. A satellite in *near-polar* orbit at about 1000 km altitude makes about 14 orbits per day and can view all parts of the atmosphere at least twice per day.

Radiation from the earth–atmosphere system reaches an orbiting satellite over a wide range of wavelengths. In the ultraviolet, visible and near infrared, solar radiation is scattered and reflected from the surface, from clouds, from aerosol (small particles suspended in the atmosphere) and from molecules. In the infrared and microwave regions, at wavelengths almost completely separated from those where solar radiation is important (fig. 2.1), radiation is emitted from the surface, clouds and molecules. Over this wide range of wavelengths a great deal of information is contained about the structure and composition of the atmosphere below.

Fig. 12.2. Rainfall display from a network of 14 weather radars (11 in UK, 2 in Eire and 1 in Jersey) at 1000Z on 5 November 1999. The envelope of the black circles shows the limits of the radar coverage. Large scale ascent due to baroclinic instability and small scale rising motion due to convection are both important in the generation of rain; there is also a high degree of organization associated with dynamical processes occurring on an intermediate scale, known as the mesoscale. The rainfall patterns, normally depicted in colour, are shown here as three shades of grey, corresponding to light (<1 mm h^{-1}), moderate ($1-8$ mm h^{-1}) and heavy (>8 mm h^{-1}) rain. Note especially the warm and cold fronts with occlusion and the pronounced line convection extending for hundreds of km near the leading edge of the cold front. (From the UK Meteorological Office)

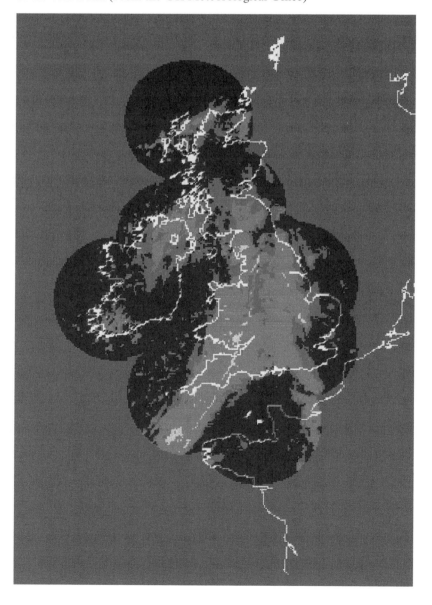

Fig. 12.3. Height profiles of mean meridional and zonal winds as measured over Aberystwyth, UK, by a MST radar for period 25–27 June 1990 shown as broken and continuous lines respectively. The standard deviation of the spread at each height over the 48-hr period is indicated by the dotted lines. Also indicated are the tropopause levels for each of the three days. (After Thomas, 1999)

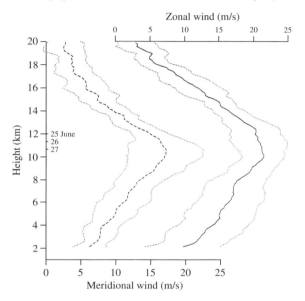

Interpretation of these remote sounding observations is often complex and difficult, but as we have seen, they possess the enormous advantage, compared with conventional and in situ observations, that a satellite can provide observations over a very large area in a short time.

The first weather satellite was launched in 1960; it carried television cameras for viewing clouds. For the first time complete pictures were seen of the patterns of clouds associated with large weather systems. Such information is now produced routinely from a large number of satellites and provides valuable information for short range weather forecasting. Detailed cloud pictures of Mars (fig. 10.15), Venus (fig. 2.7) and Jupiter (fig. 13.9) from space probes have also provided a surprisingly large amount of information about the circulation of their atmospheres.

To create images of the earth from satellite-mounted instruments it is convenient to employ simple scanning systems that generally use the motion of the satellite itself to perform some of the scanning. For instance, polar orbiting satellites are typically stabilized so that one face of the satellite always faces the earth. In the *Advanced Very High Resolution Radiometer (AVHRR)*, in front of a telescope which focuses radiation onto detectors sensitive to different parts of the spectrum, a rotating mirror is mounted at 45° to its axis of rotation (figs. 12.9 and 12.18). This scans continuously across the direction of motion of the satellite. As the satellite and the swath of observation moves forward images of different parts of the spectrum are generated. Most geostationary satellites for meteorological observation are spin-stabilized with the axis of spin parallel to the earth's axis. They carry imaging instruments called *spin-scan cameras*. Scanning in longitude is achieved by the satellite's spin; a tilting mirror at the front of the instrument arranges for scanning in latitude.

In these instruments, channels sensitive to visible radiation create images of sunlight reflected from the earth and its atmosphere. Images of radiation in the infrared emitted by the earth's surface or the atmosphere also provide important information. The most obvious information of this kind is of the temperature of the earth's surface or of the cloud top below the satellite. In atmospheric window regions, for instance between 10 and 12 μm in wavelength (fig. 12.7), the radiance received corresponds quite closely to the Planck black-body function at the surface or cloud-top temperature (fig. 12.4). Suitable small corrections have to be applied for atmospheric transparency and surface emissivity (problem 12.2). Measurements over broad spectral regions provide information about the earth's radiation budget over different areas (e.g. fig. 4.4).

By combining measurements at different wavelengths at which the properties of the surface or of clouds differ slightly, more detailed information can be retrieved. Fig. 12.5 illustrates the identification at night of the detailed distribution of fog over England and Wales by such a method.

Measurements at higher spectral resolution in different infrared or microwave parts of the spectrum provide more precise information about the atmosphere's temperature structure and composition. These will be considered in the following sections.

12.6 Remote sounding of atmospheric temperature: theory

At any frequency in the infrared or microwave parts of the spectrum where an atmospheric constituent possesses strong absorption the emitted radiation intensity from that constituent observed from above the atmosphere is a function of the distribution of the emitting gas and the distribution of temperature throughout the atmosphere.

Constituents such as carbon dioxide with strong absorption bands (fig. 12.7) at 15 μm (667 cm^{-1}) and 4.3 μm (2300 cm^{-1}) and molecular oxygen with absorption (fig. 12.10) near 5 mm wavelength (60 GHz) in the microwave region are very nearly uniformly mixed, at least up to levels

Fig. 12.4. Infrared image from the Japanese Global Meteorological Satellite GMS-1 taken at 0300 GMT on 12 October 1979. Note the tropical cyclones near the centre of the picture.

~90 km altitude. Provided that local thermodynamic equilibrium (LTE) applies (§5.7) the emitted intensity in these bands can be considered to be dependent only on the atmospheric temperature distribution.

At a given wavenumber $\tilde{\nu}$ the intensity of radiation $I_{\tilde{\nu}}$ (known as the *radiance*) received by a satellite-mounted radiometer viewing vertically downwards is given by (4.21)

$$I_{\tilde{\nu}} = \int_0^\infty B_{\tilde{\nu}}(T)\frac{d\tau_{\tilde{\nu}}(z,\infty)}{dz}dz + B_{\tilde{\nu}}(T_s)\tau_{\tilde{\nu}}(0,\infty) \qquad (12.1)$$

where T_s is the surface temperature.

Fig. 12.5. This image of the southern UK and the near continent taken at 0433 GMT on 23 October 1983 was constructed from two infrared channels of the Advanced Very High Resolution Radiometer aboard NOAA-7. The differences between the brightness temperatures viewed at $3.7\,\mu$m and $10.8\,\mu$m have been coded as shades of grey.

Objects with the same brightness temperature in the two channels such as the land and the ocean surfaces are shown as mid-grey. The brightness temperatures of thin cirrus cloud and certain small but hot sources such as gas flares are higher at $3.7\,\mu$m than at $10.8\,\mu$m. They are depicted as dark shades. The emissivity, and hence the brightness temperature, of water clouds is lower at $3.7\,\mu$m than at $10.8\,\mu$m. These appear as light shades. Low lying water cloud or fog can therefore readily be distinguished from the surface below (after Eyre et al., 1984). The marked lines at the coasts are artifacts caused by a slight mis-registration of the two channels. (Courtesy of the Meteorological Office)

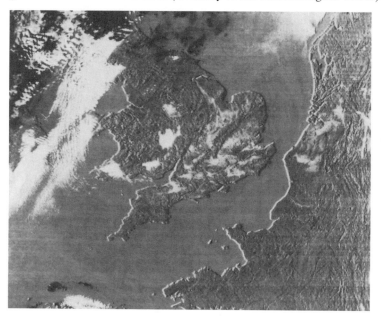

The variable $y = -\ln p$, where p is the pressure in atmospheres, is a more convenient height-dependent variable to use, so that (12.1) can be written

$$I_{\tilde{v}} = \int_{\ln p_0}^{\infty} B_{\tilde{v}}(T)\kappa(y)dy + B_{\tilde{v}}(T_s)\tau_{\tilde{v}}(0,\infty) \qquad (12.2)$$

$\kappa(y)$ is known as the *weighting function*. For a spectral region with a uniform absorption coefficient $k_{\tilde{v}}$ independent of altitude (problem 12.3 and fig. 12.6),

Fig. 12.6. 'Weighting functions' appropriate to a radiometer sounding the atmosphere by observing radiation emitted vertically upwards from the atmosphere for (a) an atmosphere with uniform absorption coefficient, problem 12.3, (b) a frequency in the wing of a pressure broadened spectral line, problem 12.4, (c) an Elsasser band, problem 12.5. Note the greater vertical resolution of (b). p_m is the pressure at which the functions peak.

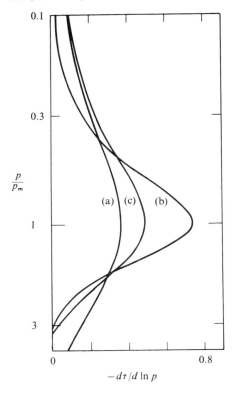

$$\kappa(y) = k_{\bar{\nu}} c p g^{-1} \exp\left(-k_{\bar{\nu}} c p g^{-1}\right) \tag{12.3}$$

where c is the mass mixing ratio of the absorbing constituent. The function $\kappa(y)$ has a peak at $k_{\bar{\nu}} c p g^{-1} = 1$, i.e. where the optical depth measured from the top of the atmosphere is unity. This weighting function is the same function that we met in §5.5 when discussing the photodissociation of oxygen in the upper atmosphere; there it was called a Chapman layer.

The weighting function describes the region of the atmosphere from which the radiation received by the radiometer originates. Note that it is broad in its height range (fig. 12.6) so that if atmospheric temperature is deduced from radiance measurements at any given frequency, it is the *average* temperature over the part of the atmosphere described by the weighting function that is being measured.

Fig. 12.7. Thermal emission from the earth plus atmosphere emitted vertically upwards and measured by the infrared interferometer spectrometer on Nimbus 4, (a) over Sahara, (b) over Mediterranean, (c) over Antarctica. The radiances of black bodies at various temperatures are superimposed. (From Hanel et al., 1971)

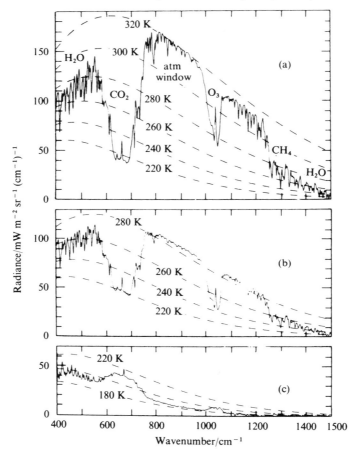

Inspection of the spectrum of infrared radiation emitted outwards by the atmosphere shown in fig. 12.7 illustrates how remote temperature sounding is achieved. In the atmospheric 'window' region between 800 and 950 cm^{-1} (10–12 μm in wavelength), in the absence of clouds, the radiation originates mostly from the earth's surface. Moving from the atmospheric window to smaller wavenumbers, the mean absorption coefficient of carbon dioxide gradually increases so that the average level in the atmosphere being monitored moves up in altitude until the most absorbing region is reached – the Q branch near 667 cm^{-1} – where the radiation largely originates in the stratosphere. For figs. 12.7(a) and (b), the tem-

Fig. 12.8. Temperature profile (full line) retrieved from measurements of radiance in the 4.3 μm and 15 μm CO_2 bands by the high resolution infrared sounder and from measurements in the 5 mm O_2 band by the scanning microwave spectrometer on Nimbus 6.

  The sounding is 48°N, 15.2°E on 23 August 1975; the circles are a nearby radiosonde. The presence of 3% of high cloud and 24% of medium cloud was deduced from the retrieval process. (After Smith, 1976)

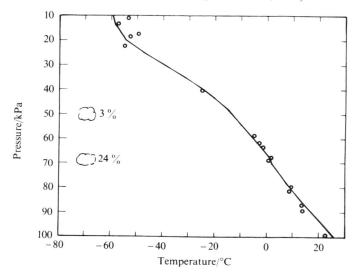

perature decreases with the altitude being monitored up to the tropopause, then increases again, while for fig. 12.7(c), observed over Antarctica, the temperature increases with altitude all the way up.

  Because of the wide range of altitude being monitored at any given frequency and because of the disturbing influence of cloud cover on the measurements, to obtain adequate observations of the atmospheric temperature structure, measurements with high accuracy from a range of frequencies from both the infrared and microwave regions need to be compounded. An advantage of microwave observations is that, because of the longer wavelengths being observed, they are less affected by the presence of cloud than those in the infrared.

  This process of the deduction of the temperature structure from such a range of observations is known as *retrieval*. In fig. 12.8 is illustrated one of the first such retrievals to be carried out from instruments on the Nimbus 6 satellite in 1976. In recent years much attention has been given to retrieval techniques which provide effectively the most relevant information for different purposes. It is particularly important to ensure that satellite remote sounding data are appropriately organized

Fig. 12.9. Schematic of the radiometer flown on the ITOS satellites (after Schwalb, 1972) showing the elements of a simple filter radiometer, including a scan mirror (which can be pointed at the atmosphere, cold space or a black body at known temperature for calibration purposes), a telescope and two spectral channels separated by a dichroic plate.

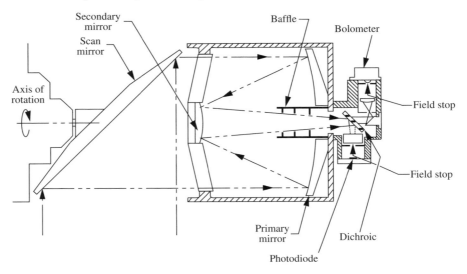

so that maximum information can be efficiently conveyed to forecasting models.

In the next sections will be described the principles of some of the instruments which have been used for remote temperature sounding and a brief account will be given of how atmospheric composition may also be deduced from spectroscopic measurements.

12.7   Instruments for remote temperature sounding

The essential elements of an infrared radiometer with which to make radiance measurements for remote sounding are illustrated in fig. 12.9. They are a telescope to focus the radiation onto a radiation detector, a filter to isolate the spectral region required, a means to calibrate the instrument with observations of 'space' away from the earth (effectively zero radiation) and with observations of a black body at known temperature. A mechanical 'chopper' may be added to interrupt rapidly the incoming signal so that it can be more easily amplified and the effective signal to noise ratio increased. For an instrument operating at microwave frequencies the elements are essentially the same with the differences that the fine spectral filtering is electronic rather than optical (the signal is mixed with the output of a local oscillator).

Reference to fig. 12.6 demonstrates the importance of high spectral resolution in the observations in order to achieve the best vertical resolution. Further, since vertical detail in the retrieval is achieved by making simultaneous observations at many frequencies for which the respective weighting functions overlap (fig. 12.10), high radiometric accuracy is also essential. In fact, an accuracy in the deduced equivalent temperature is required of the order of 0.1°C (problem 12.6).

From an orbiting spacecraft, the time available for observation of a given region of the earth's surface is determined by the size of the sample and the orbital geometry. Within this limited time, the achievements of high spectral resolution and high radiometric accuracy are very demanding and, in general, conflicting requirements. To achieve both requires instrumental ingenuity and appropriate compromises. The following paragraphs briefly describe different instruments that are currently employed for temperature sounding.

The radio technology employed in microwave measurements enables spectral resolution to be achieved of less than a typical line width with good radiometric accuracy. It therefore comes close to the ideal specified above. The Advanced Microwave Sounding Unit (AMSU) (fig. 12.18)

Fig. 12.10. (a) Atmospheric absorption spectra in the microwave region of oxygen and water vapour for a typical atmospheric zenith path showing the locations of the temperature and water vapour sounding channels of the AMSU instrument; (b) weighting functions for nadir viewing of the temperature sounding channels of the AMSU instrument as numbered in (a). (More information in NOAA KLM User's Guide, available at http://www2.ncdc.noaa.gov)

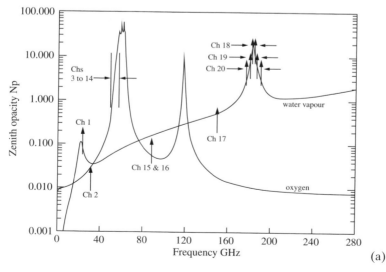

(a)

Fig. 12.10. (*continued*)

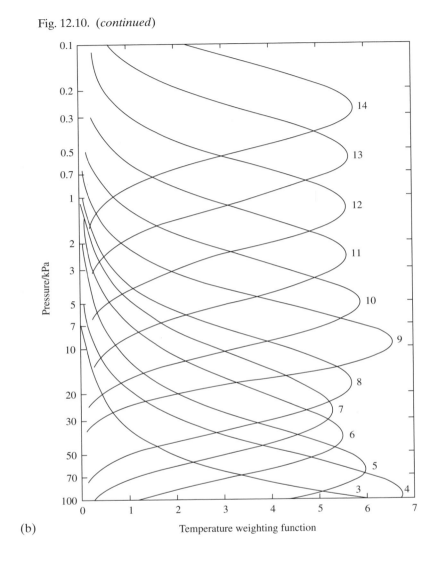

(b)   Temperature weighting function

observes at 12 separate frequencies in the oxygen band near 60 GHz (fig. 12.10(a)). The weighting functions for the various spectral channels cover the height range from the surface to about 45 km altitude (fig. 12.10(b)). The channels all lie in the wings of pressure broadened lines and in terms of vertical spread the weighting functions are therefore similar to curve (b) of fig. 12.6.

    With an infrared filter radiometer, to obtain sufficient radiometric accuracy the spectral resolution has to be significantly larger than the line width leading to weighting functions typically between curves (b) and (c)

Fig. 12.11. Schematic diagram of infrared interferometer spectrometer (IRIS) on the Nimbus 3 satellite (from Hanel et al., 1970), showing the servo-controlled moving mirror drive and the secondary interferometer receiving light from a monochromatic source for reference purposes.

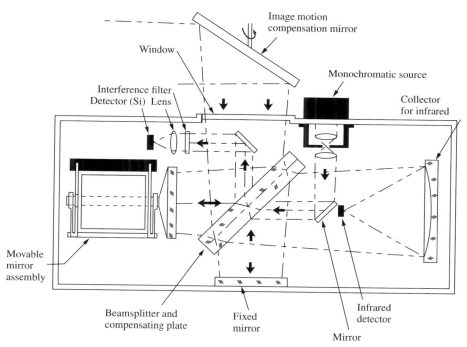

of fig. 12.6. The High Resolution Infrared Radiation Sounder (HIRS) (fig. 12.18) observes in seven spectral regions of the $15\,\mu m$ carbon dioxide band and in five spectral regions of the $4.3\,\mu m$ carbon dioxide band. The advantage of including observations in the shorter wavelength band at $4.3\,\mu m$ is that the dependence of radiance on temperature is stronger providing more information regarding the detailed structure of the warmer lower levels of the troposphere.

An instrument which collects radiant energy in a particularly efficient way while at the same time achieving high spectral resolution is the Michelson Interferometer (fig. 12.11). The incident radiation is divided within the instrument into two separate beams which are later combined. As the path difference between the two beams is changed, an interferogram is generated from which the spectrum is obtained by Fourier transformation. As the whole spectrum is observed all the time, the Michelson Interferometer possesses what is called the *multiplex* advantage; observations from many spectral channels can be made simultaneously. An early

Fig. 12.12. Pressure modulator cell as flown on the Nimbus 6 satellite in 1975 showing the method of pressure modulation (the piston mounted on springs) and the technique for changing the mean pressure in the cell, which is 6 cm long, by altering the controlled temperature of the molecular sieve. (After Curtis et al., 1974)

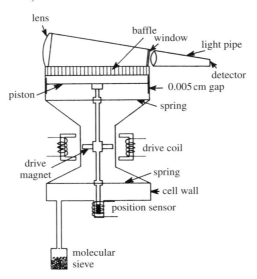

instrument, the Infrared Interferometer Spectrometer (IRIS: fig. 12.11) was used to produce the spectra shown in fig. 12.7. The Infrared Atmospheric Sounding Interferometer (IASI) is being built for the Eumetsat Polar System (fig. 12.18). It covers the wavelength range from 3.6 to 15.5 μm (2760 to 645 cm^{-1}) with a spectral resolution of about 0.25 cm^{-1}. It could therefore be said to possess over 8000 spectral channels. Because the weighting functions of these channels overlap to a large extent, the information provided by the channels is not independent. In fact studies show that it will provide about 19 independent pieces of information regarding the temperature structure and the water vapour content. This is, however, potentially about double what other infrared instruments can provide.

A less conventional way of achieving spectral selection of the radiation required is to employ the technique of *gas correlation spectroscopy*. With this technique, the gas whose atmospheric emission is being observed is employed in the instrument as its own spectral filter. In the Pressure Modulator Radiometer (PMR) as developed for atmospheric temperature sounding, a cell containing carbon dioxide is included in the instrument's optical path (fig. 12.12). The pressure of the gas in the cell is modulated by a mechanical system that results in modulation of the radiation reaching the detector only in those spectral regions where the absorption by the gas is changing. A relation exists between the mean gas pressure in the cell and the atmospheric pressure at the altitudes covered by the weighting function appropriate to the instrument. If the amount of gas in the cell is such that the absorption lines are 'strong' (see §4.3), the weighting function $\kappa(y)$ for a PMR instrument is given by (problem 12.15)

$$\kappa(y) = p^{*2}\left(1 + p^{*2}\right)^{-3/2} \tag{12.4}$$

where $p^* = ap/p_{c0}$, where p_{c0} is the mean pressure of the gas in the cell and a a constant dependent on the length of the cell and the concentration of the gas in the atmosphere. By varying the pressure of carbon dioxide in the cell, therefore, the temperature structure over the stratosphere and mesosphere, from 25 km to nearly 90 km in altitude, can be observed.

A further way in which scanning in altitude of the emitting region can be achieved is by utilizing the Doppler shift that arises between the atmospheric emission lines and the absorption lines of the gas in the cell when there is relative motion along the line of sight between the radiometer and the emitting atmosphere. Since the speed of a typical orbiting satellite is about twenty times molecular speeds at normal atmospheric temperatures, only about 5% of the satellite's velocity is required to produce a Doppler shift equal to the Doppler width of the spectral lines. By altering the direction of view from vertically downwards with no relative velocity between the satellite and the atmosphere below to about 15° from the nadir along the direction of flight, the Doppler shift can be varied and the atmospheric emission lines scanned across the absorption lines in the cell. In fig. 12.13 is shown a cross-section of atmospheric temperature structure deduced from the PMR and the Selective Chopper Radiometer (SCR – another gas correlation instrument). Stratospheric Sounding Units (SSUs) on Tiros operational satellites have provided since 1978 continuous monitoring of the temperature of the stratosphere with the PMR technique.

12.8 Remote measurements of composition

Given knowledge of the atmospheric temperature provided from remote observations on the bands of carbon dioxide or oxygen, observations of the spectrum of radiance from other emitting gases can provide information regarding their atmospheric distribution. Of especial importance is the distribution of water vapour, information about which is available from both infrared (fig. 12.7) and microwave (fig. 12.10) emission

Fig. 12.13. Temperature (K) cross-section of the atmosphere from 80 N to 80 S deduced from radiance measurements from the Selective Chopper Radiometer on Nimbus 5 and the Pressure Modulator Radiometer on Nimbus 6 for 4 August 1975.

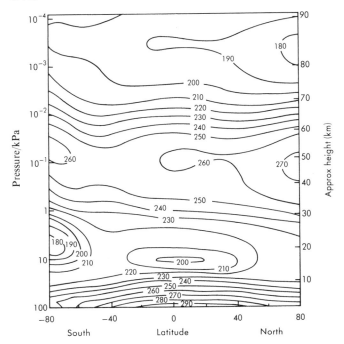

bands. The ozone distribution may also be observed from its emission band in the infrared (fig. 12.7).

Also important for remote sounding observations of ozone, particularly at high levels, is the ultraviolet region. Solar radiation backscattered from the atmosphere is strongly affected by ozone absorption in the 200 to 300 nm region. To illustrate the method, consider a simplified situation in which an ultraviolet spectrometer observes the radiation leaving the atmosphere in a vertical direction, the sun being overhead (fig. 12.14). In this spectral region and at the levels considered, attenuation of solar radiation by ozone absorption is much greater than that due to scattering (problem 12.11); further single scattering only will be considered.

At a given frequency ν the intensity of solar radiation at the top of the atmosphere is $I_{s\nu}(0)$ and at a level where the pressure is p is

$$I_{s\nu}(p) = I_{s\nu}(0)\exp\{-k_\nu n(p)\} \tag{12.5}$$

k being the absorption coefficient per molecule of ozone and $n(p)$ being the number of molecules above the level of pressure p.

Fig. 12.14

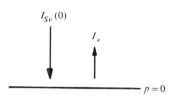

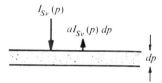

From the molecules in a layer of thickness dp at level p the radiation scattered vertically upwards will be proportional to $I_{sv}(p)$ and to dp, say it is $aI_{sv}(p)dp$ (fig. 12.14). Further attenuation will occur on traversing the return path to the satellite, so that neglecting any scattering from the lower atmosphere, the measured intensity will be

$$I_\nu = I_{sv}(0)a\int_0^\infty \exp\{-2k_\nu n(p)\}dp \qquad (12.6)$$

For purposes of illustration consider a uniform mixing ratio of ozone, i.e. $n(p) = n'p$ say, n' being a constant and

$$I_\nu = I_{sv}(0)a\int_{-\infty}^\infty p\exp(-2k_\nu n'p)d(\ln p) \qquad (12.7)$$

The quantity under the integral describes the region from which the backscattered radiation originates, i.e. the region of the atmosphere being monitored at that spectral frequency. It has a peak at $p = (2k_\nu n')^{-1}$, is very similar to the weighting function of (12.3) and has the same form as the Chapman layer of §5.5. Since the numerical value of the integral is unity, we see from equation (12.7) that the intensity of the backscattered radiation at frequency ν depends on the mixing ratio of ozone molecules and the absorption coefficient at that frequency.

By choosing a set of wavelengths for which k_ν is different, it is therefore possible to monitor the ozone concentration at different levels. Such measurements were made with the Back-Scatter Ultraviolet Spectrometer (BUV) on Nimbus 4 (Krueger et al., 1973) launched in 1970 and have been continued on the Tiros series of satellites.

Fig. 12.15. Illustrating limb sounding of the earth's atmosphere. Measurements of emission from the atmosphere's limb have the advantages of (1) a very long emitting path is viewed so that constituents present in very small concentrations can be studied, (2) near-zero radiation background beyond the limb.

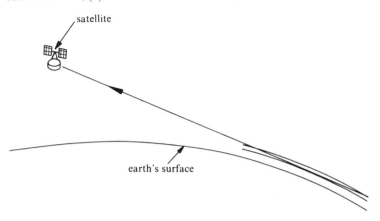

There are many other minor constituents in the atmosphere such as CH_4, N_2O, CO, NO, NO_2, HNO_3, etc., that are involved in ozone chemistry (§5.5) or pollution studies. These all possess absorption bands in the infrared or microwave regions and therefore, in principle, are amenable to remote spectroscopic observation. However, they are present in small concentrations, some only in a few parts in 10^9, so that observation of their spectral signatures when looking vertically downwards against the varying background of clouds or of the earth's surface is not easy. These problems can be overcome by observing radiation emitted (or absorbed from incident solar radiation) from the atmospheric limb (figs. 12.15 and 12.16 and problems 12.9 and 12.10). This provides a long absorbing path or a long emitting path against the cold background of space.

Many innovative instruments have been designed and built applying a variety of spectroscopic techniques to the challenges of measurement of the distribution of minor constituents in the atmosphere. Examples of the degree of innovation and sophistication that have been achieved are found in the payload of the Upper Atmosphere Research Satellite (UARS) (fig. 12.16) launched in 1991 to observe details of stratospheric and mesospheric structure, composition and dynamics.

In addition to measurements of gaseous composition it is also important to observe the distribution of liquid water in the atmosphere and the distribution of precipitation. Because of the varying spectral properties of water vapour and liquid water in the microwave part of the

Fig. 12.16. The Upper Atmosphere Research Satellite (UARS) launched by NASA in 1991 carries a number of limb viewing remote sounding instruments for observing atmospheric structure and composition. Many of the instruments were particularly directed to measuring the distribution of chlorine and nitrogen compounds involved in ozone chemistry (§5.6). Three instruments observe emission of radiation from minor constituents in order to determine their distribution, namely the Cryogenic Limb Array Etalon Spectrometer (CLAES) which employs a scanning cooled infrared spectrometer, the Improved Stratospheric and Mesospheric Sounder (ISAMS) which employs a number of pressure modulator radiometers to select radiation from different constituents, and the Microwave Limb Sounder (MLS) which observes emissions in the microwave region of the spectrum. The Halogen Occultation Experiment (HALOE) observed infrared absorption by minor constituents as solar radiation was occulted by the atmosphere. Two instruments achieved very high spectral resolution and observed the Doppler shifts of absorption or emission lines in order to measure wind velocities in the upper atmosphere. The High Resolution Doppler Imager (HRDI) included a Fabry-Perot Interferometer and the Wind Imaging Interferometer (WINDII), a Michelson Interferometer to attain the spectral resolution required. Other instruments measure solar radiation and solar particles. The Multimission Spacecraft (MMS) and the Tracking and Data Relay (TDRS) antenna are also shown. UARS is a large satellite about 9 m long and weighing about 5 tons.

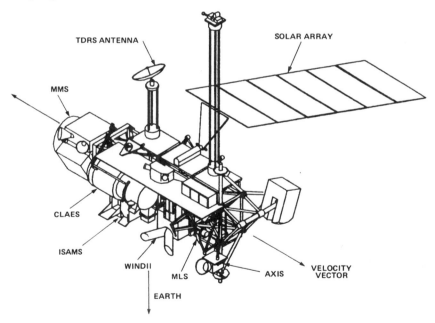

spectrum, observations near 1.5 cm wavelength can assist in providing such information especially over the oceans.

12.9   Other remote sounding observations from space

Because of the breakdown of the geostrophic approximation in tropical regions, good measurements of the wind field are required, at least at low latitudes. Observations of the motion of suitable clouds using images from geostationary satellites provide good information at two levels in the troposphere, namely near 90 kPa where small convective clouds can be tracked and near 20 kPa where cirrus clouds occur. Such observations are, of course, limited to regions where suitable clouds are present.

Observations of the state of the land or sea surface are also important to provide for general circulation models appropriate information regarding the lower boundary especially that relevant to the exchange of momentum, heat and water vapour across that boundary. Accurate measurements of surface temperature especially over the oceans requires careful radiometry at infrared wavelengths (problem 12.2). Imagery at microwave frequencies can be interpreted to distinguish between water and ice and also to provide information about surface wetness and vegetation cover.

Conditions at the surface of the earth can also be observed from the character of backscattered signals from satellite-borne radars. The speed and direction of the surface wind over the oceans can be measured by a scatterometer observing the radar returns, at a wavelength of a few cm, from capillary waves on the sea surface.

New possibilities for remote sensing arise from the exploitation of the Global Positioning System (GPS) which employs a set of 24 orbiting satellites to provide for extremely precise navigation. Signals from these satellites can also be employed for the determination of the atmospheric refractive index at the satellite frequencies which is dependent on the detailed distribution of temperature and water vapour along the observation path between the satellite and a monitor on the earth's surface (cf. problem 12.12).

Remote sounding from instruments carried on space probes is also an important means of investigating the structure of planetary atmospheres. The structure and circulation of the atmospheres of Venus (cf. problem 12.14 and figs. 12.16 and 7.10), Mars and Jupiter have been explored in this way. Further references to remote sounding instrumentation and observations from satellites and space probes are given in the bibliography.

12.10 Observations from remote platforms

To deduce the atmospheric density field, in addition to measurements of temperature, the pressure at a reference surface is required. A number of possible ways have been put forward by which satellite remote sounding of surface pressure might be achieved (e.g. by observing absorption of solar radiation reflected back from the surface in the vicinity of the visible oxygen absorption bands) but none can provide the accuracy which is necessary.

However, satellites can provide an effective communication link from remote platforms. For instance free-floating buoys in the oceans with appropriate instrumentation can be interrogated regularly from orbiting satellites and can provide observations of surface pressure and other in situ measurements. Satellite links also provide for automatic communication with remote observation platforms on land and with commercial aircraft.

12.11 Achieving global coverage

Illustrated in fig. 12.17 is the deployment of polar orbiting satellites and five geostationary satellites to provide adequate coverage of the globe from space for meteorological research and predictions. Polar orbiting satellites carrying infrared and microwave remote sounding instruments have been flown in the Tiros series of the National Oceanic and Atmospheric Administration (NOAA) of the USA since the 1970s. They are flown in pairs crossing the equator at different local times so as to provide good coverage each day. A polar orbiting satellite (EPS) to be provided by Eumetsat, the organization for European Meteorological Satellites, for launch in 2002 is illustrated in fig. 12.18. It carries many of the same instruments that have been flown on the Tiros series together with some more advanced operational instruments.

The data available each day from these satellites together with conventional observations add up to a large amount (fig. 12.19). This global system of observation known as World Weather Watch (WWW) is coordinated internationally by the World Meteorological Organization (WMO) and is being continually improved as the quality of observations improves and as other techniques for observations are developed. A dedicated international communication network, the Global Transmission System (GTS), arranges for the rapid exchange of meteorological data between national meteorological services worldwide and their timely availability. Their assimilation into models is a complex process (§11.5) that requires to take into account the particular characteristics (including the error characteristics) of each type of measurement.

Fig. 12.17. Illustrating the satellite observing system for World Weather Watch; five geostationary satellites and two or more polar orbiting satellites. Remote sounding observations are made from the satellites; they are also employed as communication links to relay observations from ships, buoys, aircraft and remote observation stations.

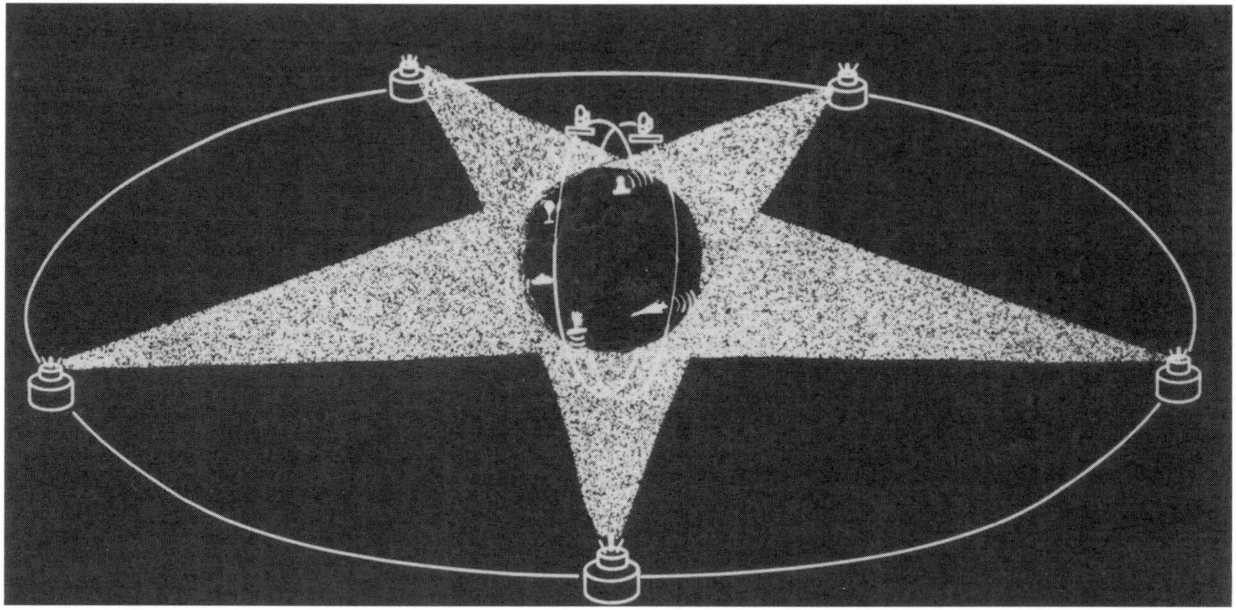

Fig. 12.18. The Metop spacecraft of the Eumetsat Polar System (EPS) due for launch in 2002. The Advanced Very High Resolution Radiometer (AVHRR) for global imagery and instruments for temperature sounding namely the High Resolution Infrared Radiation Sounder (HIRS), the Infrared Atmospheric Sounding Interferometer (IASI) and the Advanced Microwave Sounding Unit-A (AMSU-A) have been mentioned in the text. The Microwave Humidity Sounder (MHS) senses the water vapour distribution (fig. 12.10(a) shows the relevant spectrum). The Advanced Scatterometer (ASCAT) is a radar for observing surface wind over the oceans (see §12.9). The Global Ozone Monitoring Experiment (GOME) measures profiles of ozone and other atmospheric constituents using observations of scattered ultraviolet and visible radiation (see §12.8). The Global Navigation Satellite System Receiver for Atmospheric Sounding (GRAS) receives signals from satellites of global navigation satellite systems such as the USA's GPS that have passed through the earth's atmosphere. The Doppler shifts of the received signals provide information about the temperature and water vapour structure in the stratosphere and upper troposphere.

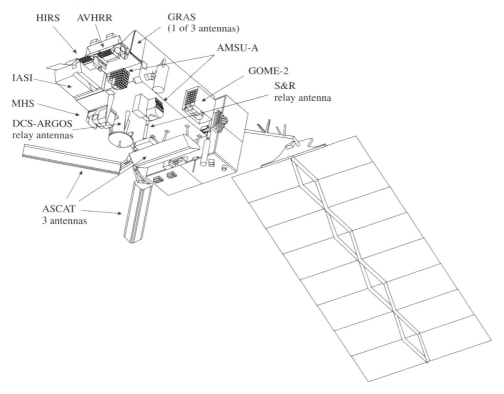

Fig. 12.19. Illustrating some of the sources of data for input into a global model at a major weather forecasting centre on a typical day (in fact for the UK Meteorological Office at 1200, 1 July 1990). Surface observations are from land observing stations (attended and unattended), from ships and from buoys. Radiosonde balloons make observations up to 30 km altitude from land and from ship-borne stations. Satellite soundings are of temperature and humidity at different atmospheric levels deduced from observations of infrared or microwave radiation. Satellite cloud-track winds are derived from observations of the motion of clouds in images from geostationary satellites. Further important routine observations in the atmosphere are made from commercial aircraft.

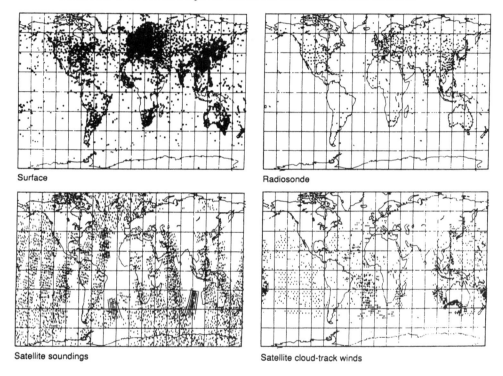

Surface

Radiosonde

Satellite soundings

Satellite cloud-track winds

Problems

12.1 The density near the 10 kPa level is deduced from surface pressure measurement combined with temperature measurements throughout the rest of the atmosphere. What is the percentage error in density resulting from (a) a 1 K error in temperature at 10 kPa, (b) a 1 K error in temperature at all levels below 10 kPa, (c) a 3% error in pressure measurement at the surface?

12.2 From the information contained in problem 4.17, and assuming the water vapour mixing ratio at any altitude is proportional to p^3

where p is the atmospheric pressure in atm, calculate the transmission of a vertical column of atmosphere in the $11\,\mu m$ window.

If $B(p)$ is the black-body function at $11\,\mu m$ appropriate to the temperature at the level where the pressure is p atm, for an atmospheric lapse rate, $B(p)$ is approximately equal to $B_0 p$ where B_0 is the black-body function at the surface temperature. Using this expression and (4.21) calculate the intensity leaving the top of the atmosphere at $11\,\mu m$. Hence estimate the error in the surface temperature deduced from a satellite observation if atmospheric absorption is ignored.

The Along Track Scanning Radiometer (ATSR), on the ERS-1 satellite of the European Space Agency launched in 1991, provides for direct corrections to be made for atmospheric absorption. Comparison is made between radiance measurements of the same area on the Earth's surface viewed from different points along the satellite's track and hence from different angles and with different amounts of absorption.

12.3 Obtain the expression in (12.3) for the 'weighting function' for an atmosphere with uniform absorption coefficient $k_{\bar{\nu}}$.

12.4 Show that in the wing of a collision broadened line of strength s, and width γ, the absorption coefficient is given by (cf. (4.5))

$$k_{\bar{\nu}} = \frac{s\gamma}{\pi(\nu - \nu_0)^2}$$

Hence show that the weighting function $\kappa(y)$ appropriate to such a frequency is of the form

$$\kappa(y) = 2\left(\frac{p}{p_m}\right)^2 \exp\left[-\left(\frac{p}{p_m}\right)^2\right]$$

where $p_m^2 = \dfrac{2g\pi p_0(\nu - \nu_0)^2}{s\gamma_0 c}$

(quantities as defined in (4.7) and (12.3)).

12.5 A useful model of a set of absorption lines is that due to Elsasser which assumes uniform line strength and spacing. For this model for collision broadened lines the transmission of a path between the level at pressure p and the top of the atmosphere is given by

$$\tau = 1 - \frac{2}{\pi^{1/2}} \int_0^{\beta p} \exp(-x^2)\,dx$$

where β depends on line strengths, widths, spacing and absorber concentration. Show that, for this case, the weighting function is given by

$$\kappa(y) = \left(\frac{2}{\pi}\right)^{1/2} \frac{p}{p_m} \exp\left(-\frac{p^2}{2p_m^2}\right)$$

12.6 Calculate from the Planck function for a source at 240 K the percentage accuracy in radiance measurements which is required to achieve a 1 K accuracy in temperature for wavelengths of (a) 4.3 μm, (b) 15 μm, (c) 5 mm.

  Refer to fig. 12.10(b) and, by estimating the degree of overlap between the weighting functions, estimate the increase in accuracy required in measurements from the individual channels if an accuracy of 1 K is to be achieved in the average temperature over layers approximately 5 km thick (approximately the difference between weighting function peaks).

12.7 A radiometer observing the atmosphere possesses an optical system with a collecting area A, field of view Ω steradians, spectral bandwidth $\Delta\tilde{\nu}$ cenred at wavenumber $\tilde{\nu}$, and mean transmission τ_0. Incident radiation from a black body of radiance $B_{\tilde{\nu}}$ is chopped sinusoidally. Show that the signal : noise ratio for measurement in t seconds is

$$\frac{S}{N} = \frac{B_{\tilde{\nu}} A \Omega \Delta\tilde{\nu} \tau_0 t^{1/2}}{2^{3/2} (\text{NEP})}$$

where NEP is the noise equivalent power of the detector (i.e. the r.m.s. radiation power incident on the detector which gives a signal equal to the noise in a 1 Hz bandwidth).

12.8 The noise equivalent temperature (NET) of a radiometer is the change of source temperature which causes a change in signal just equal to the noise. Calculate for a source temperature of 240 K the NET for an instrument with the following characteristics:

$A = 100\,\text{cm}^2$, $\Omega = 10^{-4}\,\text{sr}$, $\tau_0 = 0.5$,
$\Delta\tilde{\nu} = 1\,\text{cm}^{-1}$ at $667\,\text{cm}^{-1}$, $\text{NEP} = 10^{-10}\,\text{W}$,
$t = 4\,\text{s}$.

12.9 A satellite-mounted radiometer is observing the limb of the atmosphere. Show that the path of atmosphere traversed through the limb (i.e. $\int \rho\, dx$ where x is along the line of sight) is approximately 70 times that in a vertical path above the tangent point of the path.

12.10 Show that for the path of problem 12.9 the mean pressure along the path as defined by (4.15) is $2^{-1/2}p_0$ where p_0 is the highest pressure along the path. What is the horizontal distance between points where the pressure is $2^{-1/2}p_0$?

12.11 From the formula in problem 6.14, for a path near 60 km altitude and 0.27 μm wavelength compare attenuation due to Rayleigh scattering with that due to ozone absorption (ozone concentration at 60 km, 10^{10} cm^{-3}, ozone molecular absorption cross-section, 10^{-17} cm^2, other information in the appendices).

12.12 In an atmosphere with spherical symmetry where density ρ and hence refractive index n varies with distance r from the centre of the sphere, for a ray passing through the atmosphere the quantity nd is a constant where d is the length of the perpendicular from the centre of the sphere to the tangent to the path at the point where the refractive index is n. From this expression determine the actual position of the lowest point of a ray passing through the atmosphere to have (a) 10 km, (b) 20 km, as its lowest point.

12.13 A spacecraft emitting radio signals is occulted by the planet Venus. By studying the time taken for reception of the signals it is possible to trace the paths of rays through the Venus atmosphere during occultation and so to deduce the variation of density with altitude. Find the highest pressure on the Venus atmosphere from which such rays will emerge. (Refractive index of CO_2 at STP for radio frequencies = 1.00049.)

12.14 Suppose the AMSU instrument (weighting functions for the earth's atmosphere in fig. 12.10(b) covering the range from 0.1 to 100 kPa) were employed to view the atmosphere of Venus. From equation (12.3) and data in table 1.1 calculate the range of pressure within the Venus atmosphere which could be observed. In fig. 12.20 vertical temperature profiles for the earth and for Venus are compared.

12.15 Consider the performance of a PMR (§12.6 and fig. 12.12) as follows. For a cell 6 cm long containing pure carbon dioxide, plot out the spectrum of the absorption by a single spectral line for a line of strength 2 cm^{-1} (atm-cm)$^{-1}$, for pressures of carbon dioxide in the cell of 70 Pa and 200 Pa. Also plot out the difference between the two spectra which, assuming modulation occurs between those two pressures, is the effective transmission of the cell with respect to the modulated absorption. Also plot out the spectra, and their difference, for pressures in the cell of 4 Pa and 16 Pa. Note that the

Fig. 12.20. Vertical profiles of temperature versus pressure at 30N latitude as measured on Venus by the VORTEX experiment and on earth by the Nimbus 7 Stratospheric and Mesospheric sounder. Further details of both instruments in Houghton, Taylor & Rodgers (1984).

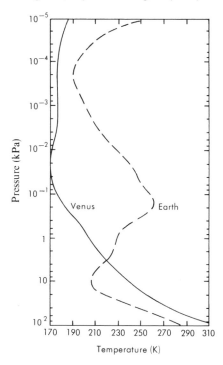

absorption for the first case is in the 'strong' region and in the second in the 'weak' region.

The weighting function $\kappa(y)$ appropriate to the PMR is

$$\kappa(y) = \frac{d\tau_\omega(y)}{dy} = \frac{d}{dy}\int_0^\infty \{\tau_\omega(\text{optics}) \times \tau_\omega(\text{cell}) \times \tau_\omega(\text{atm})\}dv$$

where $y = -\ln p$, τ_ω is the transmission of the path from the level at pressure p to the satellite made up of components from the atmosphere, the cell and the optics. The subscript ω implies that it is the part varying with the modulation frequency ω of the pressure in the cell which is selected (Taylor et al., 1972). By inserting expressions for the atmospheric and the cell transmission, assuming that in both cases the absorption is in the 'strong' region (§4.3), show that the PMR weighting function is given by equation (12.4).

13

Chaos and atmospheric predictability

13.1 Chaotic behaviour

The last few chapters have been concerned with the various forms of motion which occur within the atmosphere on a wide range of different scales, how they are observed and how they are modelled. Some of them appear to be rather stable, others show varying degrees of instability. Over the last few decades, a science of 'chaos' has developed which addresses in a fundamental way the predictability of different dynamical situations.

For a long time Newtonian dynamics appeared to be highly deterministic and therefore in principle predictable. This predictability was first questioned around 1903 by the mathematician Jules Henri Poincaré who, in investigating approximate solutions for the motion of bodies under Newtonian dynamics, found that some solutions failed to converge. He realized that the actual solutions in such cases must be highly dependent on the initial conditions, so much so that practical predictability is no longer possible.

Unpredictable situations of the kind first recognized by Poincaré, in which from infinitesimally different starting points, systems can realize very different outcomes, have become known as *chaotic behaviour*. With the advent of the electronic computer in the 1960s it has been possible to apply Newtonian dynamics to many complex situations and it has been realized that chaotic behaviour is widespread in nature. For instance, if we consider a perfect billiard table with perfectly smooth and elastic cushions and perfect billiard balls, if, after a given stroke by the cue the positions of the balls are to be predicted with reasonable accuracy after one minute of motion, the direction and magnitude of the impulse given by the cue must be known to an accuracy which would allow for all forces larger than the gravitational attraction of an electron at the opposite side of the galaxy!

Fig. 13.1. (a) Illustrating a simple pendulum consisting of a bob at the end of a string of length 10 cm attached to a point of suspension which is moved with a linear oscillatory forcing motion at frequencies near the pendulum's resonance frequency f_0. The lower diagrams show plots of the bob's motion on a horizontal plane, the scale being in cm. (b) For a forcing frequency just above f_0 the motion of the bob settles down to a simple, regular pattern. (c) For a forcing frequency just below f_0 the bob shows 'chaotic' motion (although contained within a given region) that varies randomly and discontinuously as a function of the initial conditions. (After Lighthill, 1986)

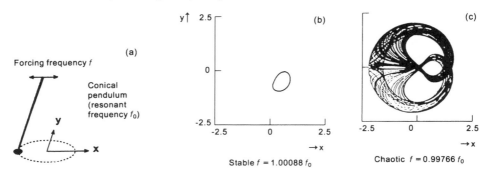

Chaotic behaviour can be exhibited by relatively simple systems, for instance in the motion of a conical pendulum (fig. 13.1). The main condition for chaotic behaviour to occur under some conditions is that the equations governing the motion in question be non-linear (problem 13.1).

For a given system, the time before substantial divergence occurs between two states beginning from closely similar initial conditions is known as the *preditability horizon*. If the initial conditions are more accurately defined the predictability horizon will move further away, but not much so. Roughly speaking, as the number of decimal places increases in the definition of the initial conditions the predictability horizon changes approximately linearly. For instance, if 4 decimal places accuracy (i.e. 0.01% error) in the initial conditions enabled prediction for 1 minute, 12 decimal places would provide for 3-minute prediction. This is because a property of chaotic systems is that neighbouring solutions diverge approximately exponentially from one another.

13.2   The Lorenz attractor

It was when studying the behaviour of an extremely simplified simulation of atmospheric motion consisting of a set of non-linear equations with just three variables (x, y and z) that Edward Lorenz at the Massachusetts Institute of Technology in 1963 stumbled across chaotic solutions. He was working with an early computer with a speed of only 60 operations per second. He restarted a particular computer integration by typing in as initial values of the variables the output of the previous run, though with fewer decimal places. The results deviated markedly from those he had obtained earlier for that integration and demonstrated to Lorenz the sensitivity of the results of the integration, after only a short time, to the detailed initial state from which the integration started. Lorenz had for the first time demonstrated chaotic behaviour in a model of the atmosphere (Lorenz, 1963). It is sometimes colourfully described as the 'butterfly effect', taken from the title of a lecture given by Lorenz in 1979: 'Predictability: does the flap of a butterfly's wings in Brazil set off a tornado in Texas?'.

The Lorenz attractor (a cross-section of it is shown in fig. 13.2) describes the complete range of the evolution of the three variables in phase space starting from different initial values. Fig. 13.2 shows that some trajectories in the attractor which start from nearby locations remain close together for a substantial time; other neighbouring trajectories diverge rapidly and end up in very different regions of phase space. Some trajectories are therefore more chaotic, and less predictable, than others.

It is clear from atmospheric studies that the real atmosphere behaves in a similar way. Some situations and circulation patterns show much more predictability than others.

13.3 Model predictability

At the end of chapter 11 (fig. 11.10) we described the improvement in forecasting skill brought about by the improvement in forecasting models. A natural question to ask is whether the improvement can continue. In other words, are there limits to the predictability of the motion and state of the atmosphere?

Using the general circulation models of the atmosphere described in chapter 11, the predictability of the model atmosphere's circulation from any given initial state can be investigated. By introducing small 'errors' into the initial parameters the variation of predictability can be found for different 'errors' in the initial state. Such experiments have been carried out with many forecasting models.

A particular example from 27 integrations over 30 days with the UK Meteorological Office global forecasting model (which at the time possessed a horizontal resolution of about 200 km and 15 levels in the vertical) is shown in fig. 13.3. To help in interpreting the diagram, consider two integrations of the model from analyses of the atmospheric state 6 hours apart. If we make the assumption that the later analysis is the true atmospheric state at its initial time, then the 6-hour forecast from the earlier

Fig. 13.2. Cross-sections (in the x, y plane) across the Lorenz attractor showing the development with time of solutions to the equations beginning from different ensembles of initial states. Parts (a) to (c) illustrate increasing divergence of solutions. (After Palmer, 1999)

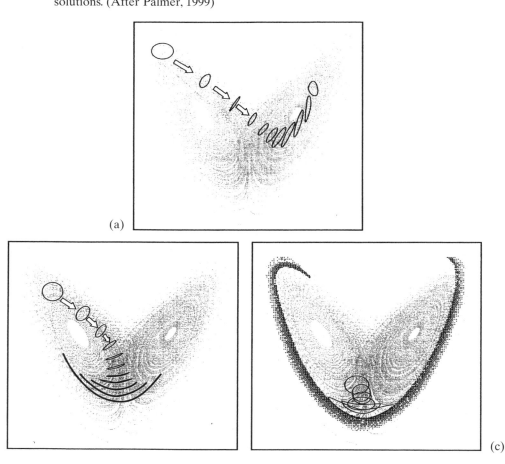

(a)

(b)

(c)

analysis can be considered as this true state plus a small error. A comparison of the integrations from these two initial states provides information on the growth of this error with time. Similar comparisons are made between runs 12 hours, 24 hours, up to 48 hours apart to find out about the growth of larger initial error fields. As the forecast period increases the rms difference (error) between any pair of runs increases until it reaches a saturation value which is a typical rms difference between any pair of model forecasts chosen at random. This represents the theoretical limit of predictability where error growth arises solely from model internal dynamics and error growth due to model imperfections is ignored.

Fig. 13.3. Upper bound on predictability. Curves show the rms differences between UK Meteorological Office model forecasts of the northern hemisphere 500 hPa height field separated by 6, 12, 18, . . . 48 hours. The curves are the averages over 27 model runs. (*a*) is an estimate of the growth of the rms differences for infinitesimal initial data error. Also plotted are the standard deviations of the model fields (model s.d.) and of real analyses of the 500 hPa height field (climate s.d.) and also the saturation level determined from the average rms difference between random forecasts. (After S. Milton: see Houghton, 1991)

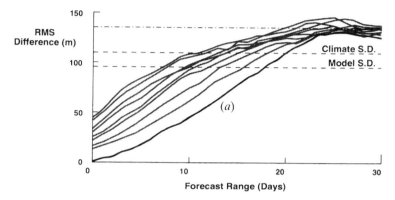

Since there appears to be approximately a linear relationship between the initial error and the limit of predictability, extrapolation can be made to estimate a limit of predictability appropriate to infinitesimal initial error. This is also shown in fig. 13.3; for this particular model it turns out to be about 26 days.

Supposing now the model atmosphere is compared with the real atmosphere by comparing atmospheric analyses with model forecasts. Fig. 13.4 shows this for the same 27 cases as in fig. 13.3. If the limit of useful forecasts is defined as that when the rms error in the forecast is equal to the standard deviation of model analyses (just under 100 m) then forecasts with this model are useful at about 6 days. If we assume that the limit of predictability of the model is a useful guide to the upper limit of the predictability of the atmosphere then the other curves in fig. 13.4 suggest that the existence of a perfect model would improve the range of useful forecasts from 6 to about 12 days; the provision of near-perfect data would move the range out to about 18 days. The ultimate range of useful deterministic forecasts appears therefore to be between 2 and 3 weeks. For this to be achieved there will need to be large improvements in data accuracy and coverage (see chapter 12) and in models (see chapter 11).

Fig. 13.4. Potential skill improvements. Curve (*c*) is the rms error of forecasts of the 500 hPa height field from a global model of the UK Meteorological Office. Curve (a) is the same as (*a*) of fig. 13.3. Curve (*b*) is an estimate (using the data of fig. 13.3) of the growth of rms difference for a day 1 rms difference equal to the day 1 rms error. (After S. Milton: see Houghton, 1991)

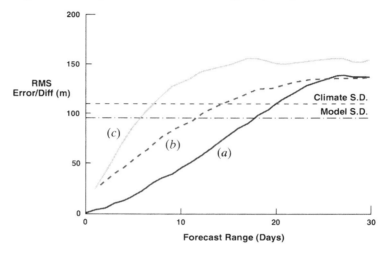

13.4 Variations in forecast skill

The results shown in figs. 13.3 and 13.4 are averages over the winter northern hemisphere north of latitude 30. But individual forecasts vary a great deal in their skill. For instance, because the large scale flow is predictable over a longer period than the small scale, forecast skill at mid latitudes is highest during winter when the large scale quasi-stationary waves are of high amplitude and lowest during the summer when smaller scale modes prevail. That some regions over particular periods show much more variability than others is easily illustrated by comparing the variability over two successive months for different regions of the northern hemisphere (fig. 13.5). Some of this difference in variability arises because of large variations in space and time of the incidence of large quasi-stationary waves known as 'blocking' situations associated with persistent areas of high pressure (fig. 13.6). Further variations in forecast skill arise because of the influence of tropical disturbances at mid latitudes. Improving the description in the model of the tropical circulation can lead to improvements of forecast skill at mid latitudes.

13.5 Ensemble forecasting

The predictability of a particular forecast can be investigated by the technique of ensemble forecasting, in which an ensemble of forecasts

Fig. 13.5. Root mean square variability in daily values of 500 hPa height in dm averaged over (*a*) January 1988, (*b*) February 1988.

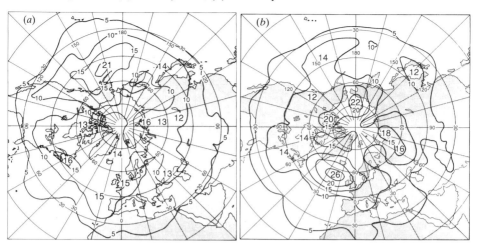

Fig. 13.6. Height in dm of the 500 hPa surface for the northern hemisphere on 7 January 1985 illustrating a strong blocking situation over the Atlantic Ocean.

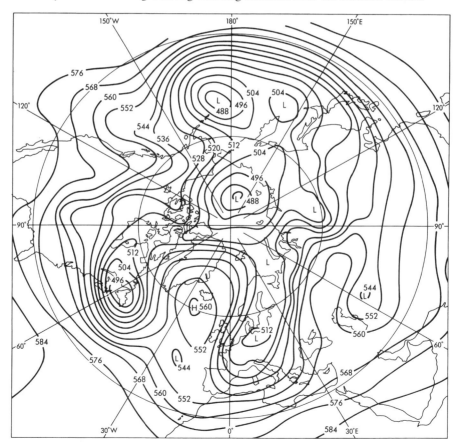

Fig. 13.7. Schematic projections of ensembles of long-range forecasts. From close initial states, some ensemble forecasts after 30 days show low overall spread, whereas others show large spread.

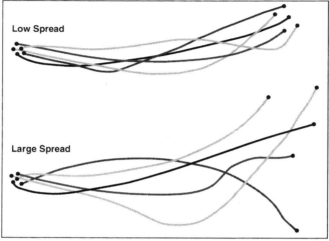

is run from a cluster of initial states which are generated by adding to the analysed initial states small perturbations within the range of observation and analysis errors. In practice, the simplest way to create the ensemble is to use forecasts from a number of consecutive analyses each separated by a constant time interval (e.g. 6 hours as for the data from which fig. 13.3 was generated) with the latest corresponding to the start of the forecast period.

The forecast provided by the mean of such an ensemble shows a significant improvement in skill compared with the individual forecasts from members of the ensemble. The time range of useful forecasts can thereby be extended. Further useful information from the ensemble forecast is an indication of forecast skill. If all the forecasts within the ensemble fall closely together (low spread, fig. 13.7), the forecast possesses higher skill than if there is high spread in the results from the members of the ensemble (further information in Palmer, 2000).

13.6 Model improvements

To achieve higher skill in forecasting, the importance of improvements in the accuracy and coverage of data has been emphasized. Improvements in the models are also important. For instance, the continuing increase in model resolution which has been possible through the availability of higher performance computers and the developments of improved

Fig. 13.8. Surface pressure (in hPa) synoptic situation at 0000Z 16 October 1987 as an unusually intense storm approached the UK. Figures in circles are observations of surface temperature (in °C); they show the very strong gradient across the warm front.

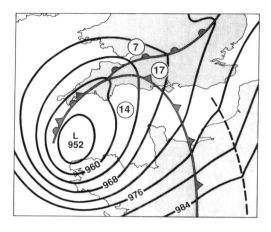

parametrizations has led to much of the increase in forecast skill shown in fig. 11.10.

A basic problem, however, remains with models of the kind described in chapter 11 which stems from the simple way in which the sub grid scale motions are parametrized. The assumption is made that the transport of momentum, heat and water vapour by the small scale motions follows Fickian diffusion (i.e. diffusion down the gradient proportional to its magnitude) with diffusion coefficients chosen to give the best results; different values may be chosen in the horizontal and the vertical, they may also vary with position and season. Further, in order to ensure numerical stability for the integrations the value of the coefficients which need to be used are typically significantly larger than are physically realistic. Since the dissipation of energy and momentum through smaller scale motions is crucial to longer term predictability, the question arises as to whether the estimates of atmospheric predictability from model studies are affected by the inadequacy of the parametrization of small scale processes.

An illustration of this inadequacy is provided by the inability of models to create or to maintain sharp air mass boundaries in the way that is common in the real atmosphere. Warm and cold fronts (see problem 7.13 and fig. 7.8) associated with mid-latitude depressions are examples of such discontinuities; a particularly intense discontinuity is illustrated in fig. 13.8). Another example is the sharp boundary which exists at the edge of the polar vortex in the southern hemisphere in the winter; air in the polar vortex does not mix significantly during the winter months with air from other parts so creating the conditions necessary for the development of the ozone hole (see §5.6) in the spring. In both these cases the model descriptions tend to blur the boundaries and generate unreal amounts of diffusion across them.

Further understanding of this problem can come from experiments with different model formulations. The kinematics in the models described in chapter 11 are set up in an Eulerian framework (i.e. the co-ordinate system is fixed with respect to the earth's surface). Another possibility is to formulate the kinematics in a Lagrangian framework (i.e. the co-ordinate system follows the flow). An example of a simple Lagrangian model as developed by Cullen and his co-workers was given in problem 11.7. Experiments by Cullen and Roulstone (1993) suggest that, because of the substantial degree of organization in small scale atmospheric motions which is not described by grid-point models, there may be less predictability present in the models than in the real atmosphere.

13.7 Jupiter's Great Red Spot

The Great Red Spot on Jupiter has been in existence at least over the three centuries since it was first observed by Robert Hooke in 1664. It is a large circulation about 15000 km in diameter and is surrounded by rapidly changing turbulent eddies; its remarkable stability continues to intrigue scientists who study fluid flow. Other similar circulations, though not so large, are also evident in Jupiter's atmosphere (fig. 13.9, and Vasavada et al., 1998). Strong dynamical similarities exist between these features in the atmosphere of Jupiter and stable, closed baroclinic eddies which can be produced in laboratory experiments on thermal convection in a rotating fluid subject to internal heating and cooling (fig. 13.9 and problem 13.3). Blocking patterns in the earth's atmosphere (see fig. 13.6) are also situations in which isolated baroclinic eddies show stability over a substantial period – perhaps several weeks. Studies relating these various phenomena are relevant to an understanding of predictability.

13.8 Different kinds of predictability

So far it is the predictability of weather on time scales of a few weeks which has been under consideration, in particular its dependence on the initial conditions. Changes in the boundary conditions, for instance changes in the distribution of solar radiation because of progress in the seasonal cycle or changes in the temperature of the ocean surface, have also been allowed for in the models. Such predictability which is depen-

Fig. 13.9. (a) A mosaic of Voyager 1 images of Jupiter projected to show the planet as seen from below the South Pole. The Great Red Spot appears in the top right of the image. Other long-lived anticyclonic eddies appear as trains of white oval spots on lines of constant latitude. (b) and (c) Streak photographs of the circulation in a rotating laboratory annulus subject to internal heating and sidewall cooling showing regular baroclinic eddies. The value of Φ (problem 13.3) is 0.73 for the conditions in which the isolated eddy shown in (b) is formed. For (c) $\Phi = 0.54$. (After Read & Hide, 1984)

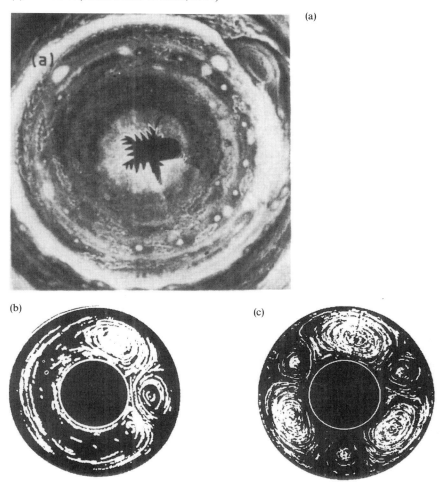

dent on the detailed initial conditions as well as on the boundary conditions has been classified by Lorenz (1975) as predictability of the *first kind*.

We are also interested in prediction on longer time scales of the average weather, or the climate. Is it, for instance, possible to predict months ahead what type of seasons are likely to occur – for instance, will they on average be warm, cold, wet or dry? Such predictions if possible will not be dependent on the initial atmospheric conditions – detailed memory of those will be lost – but they will be dependent on detailed changes in the boundary conditions. Such predictions have been classified by Lorenz as predictions of the *second kind*. It is to these I shall turn in the next chapter, which addresses questions concerned with climate change.

Problems

13.1 The existence of non-linearity in the equations of motion of a system is necessary for it to exhibit chaotic behaviour. In fig. 13.1 chaos in the motion of a conical pendulum is described; what is the source of the non-linearity in its equations of motion?

13.2 Lorenz (1982) has illustrated the problem of atmospheric predictability by investigating a simple non-linear system described by the single first order quadratic difference equation:

$$Y_{n+1} = aY_n - Y_n^2$$

where a is a constant. If $0 < a < 4$ and $0 < Y_0 < a$, a sequence is generated in which $0 < Y_n < a$ for all n.

For $a = 3.75$ and $Y_0 = 1.5$ calculate a table of Y_n for values of n up to 30. Suppose there is an error in Y_0 (representing the initial data) so that $Y_0 = 1.501$. Work out a new sequence of Y_n.

Now set Y_0 back to 1.5 and suppose that a is in error (representing an error in the model) so that $a = 3.751$. Again work out a new sequence of Y_n.

Finally to simulate the effect of errors in mathematical procedure, starting from $a = 3.75$ and $Y = 1.5$ compute a series of Y_n in which each value of Y_n is rounded off to four significant figures.

Comment on the difference between the sequences of Y_n. In particular, for each case, how many stages are required for the error to grow by a factor of 10, 100, 1000?

13.3 The character of the flow in a rotating annulus containing fluid in which thermal convection is induced depends on several dimensionless quantities. If the axial and transverse dimensions of the apparatus are H and L respectively, the rotation rate Ω, the fluid density ρ_0, the imposed density contrast $\Delta\rho$, and the kinematic viscosity ν, show that the following dimensionless parameters may be formed.

$$\Phi = \frac{gH\Delta\rho}{\Omega^2\rho_0 L^2} \quad and \quad \zeta = \frac{\Omega^2 L^5}{\nu^2 H}$$

where g is the acceleration due to gravity. When ζ is very much greater than about 10^5, the character of the flow depends mainly on Φ (Read & Hide, 1984, and figs. 10.1 and 13.9).

14

Climate and climate change

14.1 The climate system

The physical components of the climate system (fig. 14.1) include the atmosphere, the oceans, sea ice, the land and its features (including the vegetation, albedo, biomass and ecosystems), snow cover, land ice (including the semi-permanent ice sheets of Antarctica and Greenland and glaciers), and components of the hydrological cycle (including clouds, rivers, lakes, surface and subsurface water). These components all interact on a wide range of space and time scales. They also interact with components which are generally regarded as external to the system, for instance the sun and its output, the earth's rotation and the geometry of its orbit, the geographical features of the earth's surface and the mass and basic composition of the atmosphere and ocean. Detailed study of the climate involves all these components and the interactions between them.

Because of the large influence of human activities on the environment and the climate on a global scale there is a strong scientific imperative to study as fully and as carefully as possible all aspects of the likely changes and the details of the resulting impacts. International programmes such as the World Climate Research Programme (WCRP) and the International Geosphere Biosphere Programme (IGBP) have provided the means through which much of the activity has been organized and the Intergovernmental Panel on Climate Change (IPCC) has provided thorough and authoritative assessments of anthropogenic climate change.

In this chapter we shall introduce some of the fundamental processes relevant to the climate and climate change, especially those which involve the atmosphere. Understanding these fundamentals is key to understanding the impact of human activities on climate; they therefore possess a particular relevance. More details of the science of climate change can be found in Houghton (1997) and in IPCC (2001).

Fig. 14.1. Schematic view of the components of the global climate system (bold), their processes and interactions (thin arrows), and some aspects that may change (bold arrows). (From Baede et al., 2001)

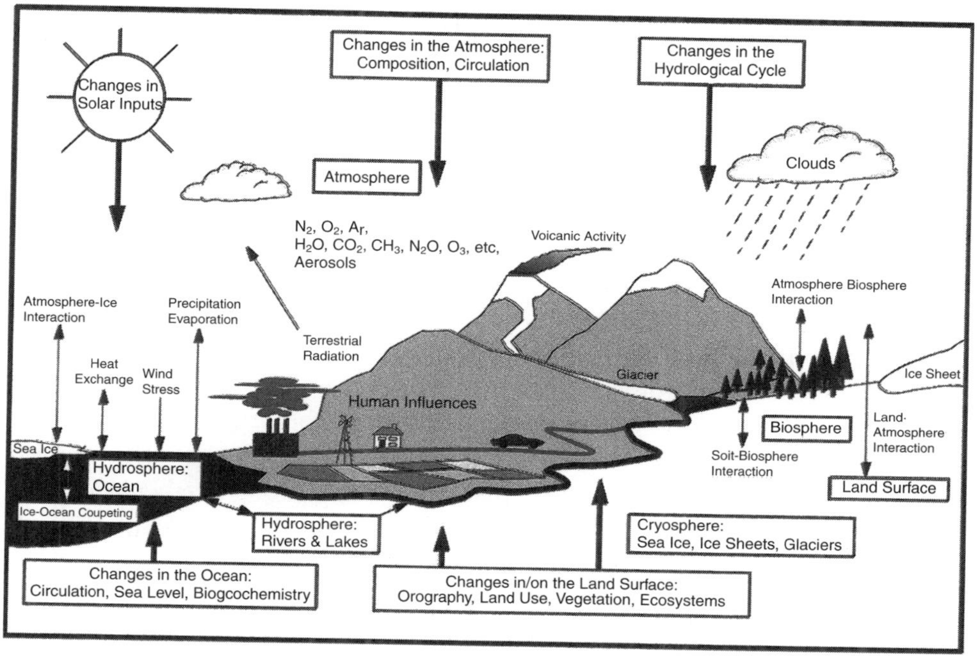

14.2 Variations of climate over the past millennium

Climate is loosely defined as averaged weather. More specifically the characters of given climate regimes are defined through the statistical properties of appropriate parameters (e.g. temperature, precipitation, solar radiation, cloudiness etc) – averaged values and deviations therefrom over regions or time scales of interest. Typical time scales may be long, concerned for instance with the study of climate over thousands of years during ice-age periods in the past, or relatively short, concerned for instance with changes over the last decade or possible changes during the next century due to human activities.

Although the term climate can refer to any part of the atmosphere or indeed the ocean, because humans live near the earth's surface, it is particularly the climate near the surface that is of interest to us. Often therefore when using the term climate, it is the climate near the surface which is implied which involves descriptions of parameters such as the near-surface air temperature and humidity, the precipitation, the wind speed and the solar radiation.

Variations of climate occur over the whole range of time scales of interest. In fig. 14.2 are illustrated the changes in the annual average of global average surface temperature over the last 140 years, which shows a lot of variability from year to year and from decade to decade together with an overall warming trend of about 0.6°C over the period. Also illustrated are changes in the average temperature of the northern hemisphere over the last thousand years which indicate a cooler period between about 1400 and 1900 known as the 'Little Ice Age'.

Several explanations have been put forward for the variations in climate over the last thousand years. First, of especial note is the strong warming at the end of the last century; globally the 1990s was the warmest decade during the instrumental period and for the northern hemisphere it was likely to have been the warmest decade of the last millennium. There is strong evidence that most of the warming during the second half of the twentieth century is due to the increase in atmospheric greenhouse gases, such as carbon dioxide (§14.5). That cannot, however, be the reason for earlier warmings, for instance that at the beginning of the twentieth century, as significant changes in greenhouse gases concentrations are relatively recent (fig. 14.7).

A second explanation is that there have been variations in volcanic activity that can inject into the atmosphere large quantities of dust and gases such as SO_2 which can lead to the formation of small sulphate particles (known as aerosol) so reducing the input to the atmosphere of solar energy. This leads to cooling, typically of half a degree C in the global average, over a period of about two years following a major eruption (fig. 14.13 illustrates the effect of the Pinatubo eruption in 1991) until the dust has been removed from the atmosphere. Such variations due to volcanic activity can be identified in the record.

A third explanation is that there could have been variations in solar radiation. Accurate direct measurements of the solar radiation incident outside the atmosphere during the last two decades are available from measurements from spacecraft; variations of about 0.1% occur during a solar cycle. Over longer periods there is indirect evidence of variations in solar radiation. For instance, it is possible that it could have been a few tenths of a per cent less during the Maunder Minimum in the seventeenth century, a period during which almost no sunspots were recorded.

A fourth explanation is that, even without any changes in external forcing, variations in climate can arise due to internal variations within the climate system especially associated with interactions between the atmosphere and the ocean and between different parts of the ocean. In fig. 14.2a

Fig. 14.2. (a) Changes in the global average surface temperature (the average of near surface air temperature over land and sea surface temperature) over the last 140 years shown as departures from the 1961–90 average, compared with simulations with a climate model in which both natural and anthropogenic forcings are included. The band of model results are for four runs of the model (from 'Summary for Policymakers', IPCC, 2001). (b) Changes in the average surface temperature of the northern hemisphere over the last millennium reconstructed from a variety of proxy data and from 1860 from thermometer data. The 95% confidence range in the annual data is represented by the grey region. (From 'Summary for Policymakers', IPCC, 2001)

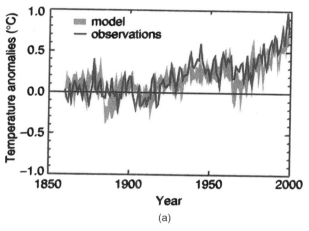

(a)

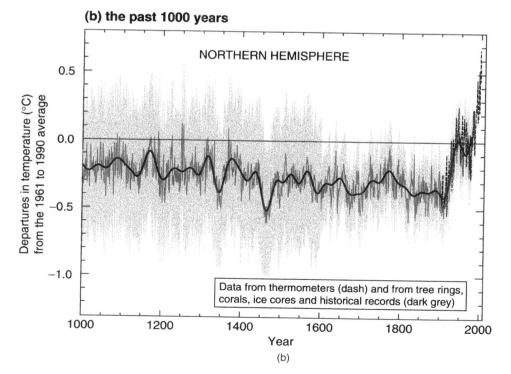

(b)

the observed record of global average temperature since 1860 is compared with simulations from climate models (§14.8) which have included in their formulation estimates of changes in greenhouse gases, aerosols and solar radiation. By including these factors the major changes on the longer time scales are well simulated. The model simulations in fig. 14.2a also illustrate internal variability that exists in the absence of any changes in external forcing and that possesses a character similar to that seen in climate observations.

14.3   The ice ages

Climatic variations from one thousand years to one million years before the present are dominated by major ice ages when ice sheets spread to cover large parts of northern Europe and North America. Data over the last quarter of a million years are available from cores of ice drilled from the ice caps over Greenland and Antarctica. Information about the temperature at which the ice was laid down can be retrieved from measurements of the ratio of the isotopes of oxygen within the ice; analyses of air trapped in bubbles within the ice provide information concerning atmospheric composition (fig. 14.3). Data are also available from sediment cores from the ocean floor.

It was first suggested in 1867 by James Croll, whose ideas were later developed by Milankovitch in 1920, that the main drivers for the ice age variations are the changes in the intensity and distribution of solar radiation reaching the earth which arise from secular changes in the earth's orbit around the sun. These are (fig. 14.4) changes in the eccentricity e of the orbit with a period of about 97 000 years, changes in the obliquity ε of its axis with a period of about 40 000 years and changes due to the precession of the longitude ω of the perihelion with a period of about 21 000 years. The changes in the distribution of solar radiation which result are particularly noticeable in the polar regions (fig. 14.5 and problem 14.3) where they amount to ±5%. Also shown in fig. 14.5 is the record of changes in global ice volume over the past million years. Study of the correlation between the global ice volume and the polar summer sunshine confirms a connection between them; 60% of the variance in the global ice volume falls close to the three 'Milankovitch' frequencies.

More careful study of the relationship between the ice ages and the earth's orbital variations shows that the size of the climate changes is larger than might be expected from forcing from the radiation changes alone. To explain the climate variations it is necessary to introduce various feedback

Fig. 14.3. Observations from the Vostok ice core, showing for the last 160 000 years the variation of atmospheric temperature over Antarctica (it is estimated that the variation of global average temperature would be about half that in the polar regions) and the atmospheric carbon dioxide concentration (after Raynaud, 1993). The range of possible values of carbon dioxide concentration during the twenty-first century is also shown.

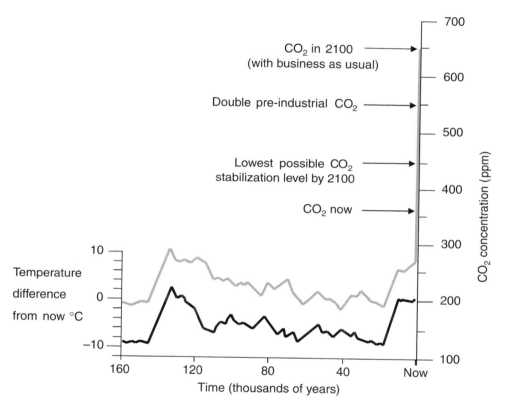

processes involving the atmosphere, ocean, ice and land and interactions between them (Berger, 1995). Of particular interest to us here are the feedbacks which occur between atmospheric temperature and the carbon dioxide concentration – carbon dioxide concentration influences atmospheric temperature through the greenhouse effect (§14.6), and carbon dioxide concentration is itself influenced by factors which depend on atmospheric temperature (fig. 14.3).

An important inference concerning the climate can be made from the degree of regularity and consistency which pertains in the climate response to the changes of solar radiation which provide the Milankovitch

Fig. 14.4. (a) Varying elements of the earth's orbit. (b) Variations of the elements during the last 250 000 years (from Berger, 1982). The eccentricity e ($= (1 - b^2/a^2)^{1/2}$ where a and b are the semi-major axis and semi-minor axis respectively) is shown dashed (left hand scale), the obliquity ε is the full line (far right hand scale) and the deviations of the precessional term ($e \sin \omega$) from its 1950 AD value are shown by the dash–dot line (right hand scale). ω is the longitude of perihelion relative to the moving vernal equinox.

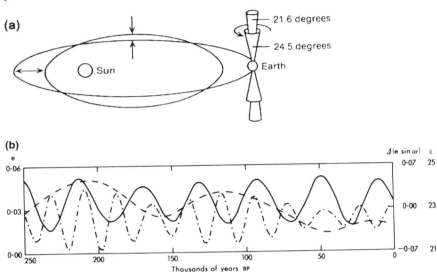

forcing. It suggests that the climate system is not strongly chaotic so far as these large variations are concerned. We can also note that changes in climate as a result of the increase of greenhouse gases are driven by changes in the radiative regime at the top of the atmosphere. These changes are not dissimilar in kind (although different in distribution) from the changes that provide the Milankovitch forcing. It can be argued therefore that increases in greenhouse gases will also result in a largely predictable climate response.

14.4   Influence of the ocean boundary

In previous chapters we have been describing the behaviour of the atmosphere through consideration of the various physical processes which pertain there or through computer models that incorporate the appropriate physics and dynamics. In this work there has been the implicit assumption that the ocean surface which underlies much of the atmosphere possesses a temperature distribution which is fixed. For the purposes of weather forecasting this is generally an adequate assump-

Fig. 14.5. The variations in the earth's orbit (fig. 14.4) cause changes in the average amount of summer sunshine (in millions of joules per square metre per day) near the poles (a), which appear as cycles in the climate record in terms of the volume of ice in the ice caps (b). (After Broecker & Denton, 1990)

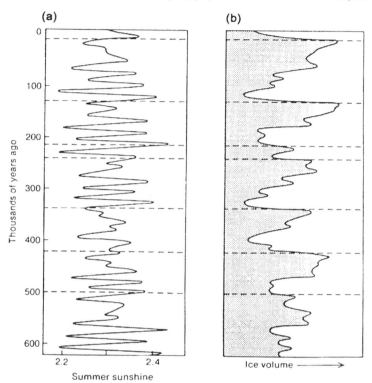

tion as the sea surface temperature changes only slowly. However, for considerations of changes over longer time scales such as those concerned with climate change, variations of sea surface temperature are very important.

It is not surprising that the atmosphere and its circulation is strongly influenced by sea surface temperature. The largest contribution to the heat input to the atmosphere (fig. 10.11) is due to evaporation from the ocean surface which, because of the rapid variation of the saturation water vapour pressure with temperature, increases rapidly with sea surface temperature. The energy which goes into that evaporation is then released in the atmosphere as latent heat of condensation.

Studies of the impact of changes in sea surface temperature on the atmosphere's circulation and on the patterns of regional climate have been carried out both from observations and climate models. These demonstrate that changes in sea surface temperature frequently influence the climate in regions far removed from where the sea surface temperature changes occur.

The largest variations in tropical sea surface temperature are found in the east tropical Pacific during periods of the El Niño when increases of sea surface temperature of up to 7 K above the normal climatological average can occur. These El Niño events occur about every five years (fig. 14.6(a)). In them, changes in the ocean circulation in the tropical Pacific are forced by changes in the surface wind, the atmospheric circulation changes in turn being driven by changes in the sea surface temperature. El Niño events can therefore only be understood or modelled by treating the atmosphere and ocean as a coupled physical and dynamical system.

Associated with El Niño events are anomalies in the circulation and rainfall in all tropical regions and also to a lesser extent at mid latitudes. Many of these anomalies are experienced as extreme events such as droughts and floods (fig. 14.6(b)). Because of these impacts much effort is going into the development of forecasting models for El Niño events.

14.5   Human influences on climate

Of particular concern to the world community in recent years has been the effect of human activities on various parts of the environment and the possible damaging impacts that might result. Local atmospheric pollution, especially in cities, has been a concern for a long time and much has been done to reduce it. But the possibility also exists for pollution on a global scale. There are two particular examples of this. One is the partial destruction of the ozone layer through the introduction of small concentrations of chlorine compounds, especially the chlorofluorocarbons (CFCs) (§5.6 and problem 5.11). The other example is that of the impact on the greenhouse effect and global climate of the emissions into the atmosphere from human activities of gases such as carbon dioxide, methane and nitrous oxide (known as *greenhouse gases* because they influence the greenhouse effect). The presence of aerosols in the atmosphere (e.g. from biomass burning or sulphates formed as a result of sulphur dioxide emissions), which can reflect or absorb solar radiation, can also have significant effect on the climate (fig. 14.11).

The most important of the greenhouse gases is carbon dioxide (cf. fig. 14.11); its growth in concentration in the atmosphere since the pre-industrial era of about 30% (fig. 14.7) has been largely due to the burning of fossil fuels (as is evidenced from carbon isotope studies) with a signifi-

Fig. 14.6. (a) Changes in sea surface temperature 1871–1998, relative to the 1961–90 average, for the eastern tropical Pacific, off Peru (courtesy Hadley Centre, UK). (b) Regions where droughts and floods occurred associated with the 1982–83 El Niño. (After Canby, 1984)

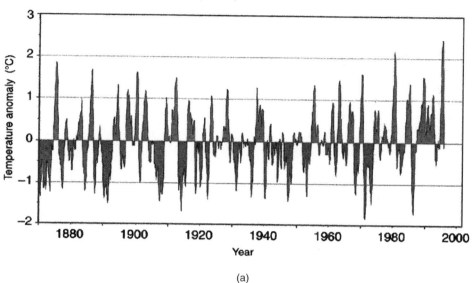

(a)

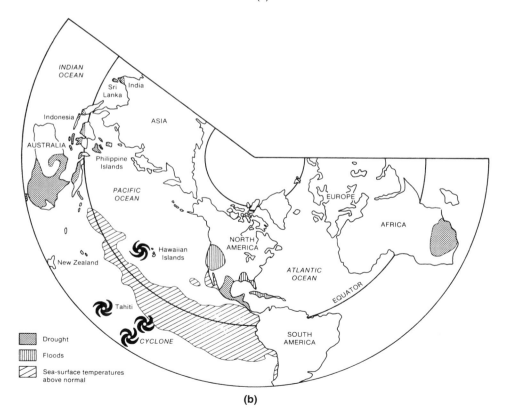

(b)

Fig. 14.7. CO_2 and CH_4 concentrations (ppm) over the past 300 years showing the rapid increase since the onset of industrialization.

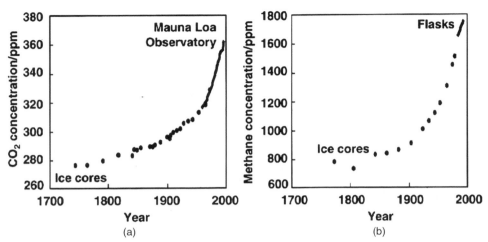

(a) (b)

cant contribution resulting from deforestation. The increases in the concentrations of methane and nitrous oxide over this period have been about 145% and 15% respectively.

Carbon dioxide provides the dominant means through which carbon is transferred in nature between a number of natural carbon reservoirs – a process known as the carbon cycle. Fig. 14.8 illustrates the way carbon cycles between the various reservoirs – the atmosphere, the oceans (including the ocean biota), the soil and the land biota. The movements of carbon in the form of carbon dioxide into and out of the atmosphere are large; about one quarter of the amount in the atmosphere is cycled in and out each year, about half with the land biota and the other half through physical and chemical processes across the ocean surface. On century time scales nearly all the anthropogenic carbon emitted into the atmosphere is not lost to these reservoirs but distributed amongst them.

The details of the distribution between the major reservoirs can be estimated from a combination of measurements of changes in atmospheric carbon dioxide and oxygen concentrations (fig. 14.9). Take up of carbon dioxide by the ocean leads to no change in atmospheric oxygen while take up by the land biosphere adds carbon to the biosphere and releases oxygen back to the atmosphere. Over the period illustrated in fig. 14.9, of the carbon dioxide emissions from fossil fuel burning approximately 40% remained in the atmosphere, the remainder being roughly equally divided between land biosphere and ocean.

Fig. 14.8. The reservoirs of carbon in the earth, the biosphere, the ocean and the atmosphere and the annual exchanges of carbon dioxide (expressed in terms of the mass of carbon it contains) between the reservoirs (from Houghton, 1997). The units are gigatonnes (Gt). Emissions of carbon dioxide into the atmosphere from the burning of fossil fuel currently amount to about 5.5 gigatonnes of carbon per annum.

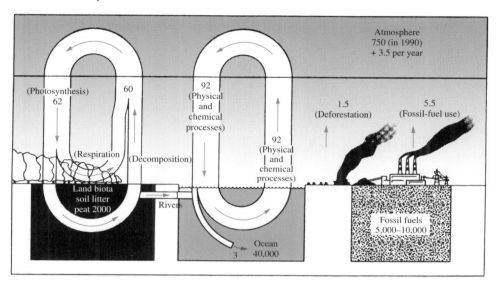

Exchanges between the reservoirs occur on a wide range of time scales determined by their respective turnover times – which range from a year to decades (for exchange with the top layers of the ocean and the land biosphere) to millenniums (with the deep ocean or long lived soil pools). Over thousands of years, before human activities became a significant disturbance, the atmospheric concentration of carbon dioxide maintained a substantially steady value implying that exchanges between the reservoirs were remarkably constant (see fig. 14.14).

The large range of turnover times means that the time taken for a perturbation in the atmospheric carbon dioxide concentration to relax back to an equilibrium cannot be described by a single time constant. Although a lifetime of about 100 years is often quoted for atmospheric carbon dioxide so as to provide some guide, use of a single time lifetime can be very misleading. Because of this long lifetime, a substantial proportion of the carbon dioxide emitted into the atmosphere from human activities today will affect its atmospheric concentration up to a century and more from now. The greenhouse gas nitrous oxide also possesses a similarly long lifetime in the atmosphere.

Fig. 14.9. The points show observations of atmospheric carbon dioxide and oxygen concentrations for the period 1990–2000. The long arrow shows the expected atmospheric trend from fossil fuel burning if there were no oceanic and terrestrial exchanges. The slopes of the lines associated with ocean and land uptake are determined by the known $O_2 : CO_2$ stoichiometric ratio of these processes. (From Prentice et al., 2001)

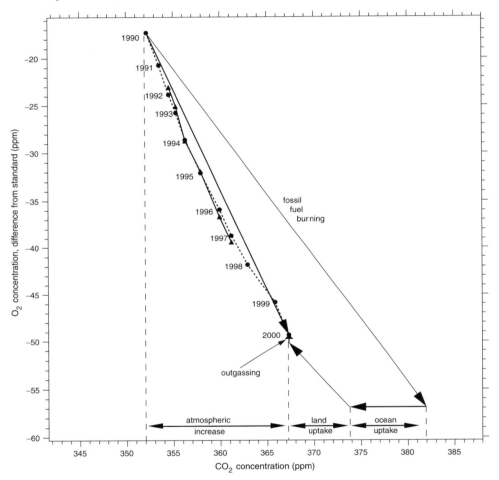

By contrast the lifetime of methane in the atmosphere before it is destroyed by chemical processes is around ten years. The lifetime of aerosols in the troposphere is very much shorter; they fall out or are washed out within a few days. In the stratosphere, micron-size particles fall out only slowly, for instance small particles that enter the stratosphere or are formed there as a result of volcanic eruptions typically remain there for about two years.

14.6   The enhanced greenhouse effect

In §1.2 the greenhouse effect was introduced with a qualitative argument showing how gases which absorb in the infrared part of the spectrum act as blankets over the earth's surface. The theory of how this occurs for an atmosphere in radiative equilibrium was developed in §2.5.

From Tables 1.1 and 1.2 we see that on the planet Venus, because of the very deep atmosphere mostly consisting of the absorbing gas carbon dioxide, the size of the greenhouse effect is very large, over 500°C. On Earth it is about 30°C; without it the average temperature of the Earth's surface would be about −15°C, well below freezing point. This is the *natural* greenhouse effect arising from the presence of the naturally occurring greenhouse gases (the most important of which are water vapour, carbon dioxide, methane, nitrous oxide and ozone) and from the presence of clouds. The absorbing bands due to most of these gases can be recognized in the spectra of radiation leaving the earth's atmosphere illustrated in fig. 12.7.

We saw in the last section that, due to human activities, the concentrations of some of the greenhouse gases are increasing thus leading to an *enhanced* greenhouse effect. Of particular importance is the increase in carbon dioxide – it has contributed about two thirds of the enhanced greenhouse effect to date. In what follows, therefore, we shall mostly illustrate the enhanced greenhouse effect from the example of carbon dioxide.

Let us suppose that the concentration of carbon dioxide in the atmosphere doubles from its pre-industrial value of about 280 ppm to 560 ppm (its value in 1998 was 365 ppm). If no strong action is taken to curb its accelerating rate of increase, this doubling will occur sometime before the year 2100. We want to work out the effect of this on the radiation fluxes in the atmosphere and the atmosphere's temperature structure. Further, because the surface temperature is probably the most important parameter describing the climate, we wish to know the effect of doubled carbon dioxide on the surface temperature.

First we look at the effect on the stratosphere. Inspection of fig. 12.7 shows that at the top of the atmosphere the radiation leaving the atmosphere from the strongly absorbing central region of the carbon dioxide band centred at $15\,\mu m$ wavelength ($667\,cm^{-1}$) originates in the stratosphere and the mesosphere. This is a region that, to a first approximation, is in radiative equilibrium. Under this assumption, in §4.8 we balanced the heating due to the absorption of solar radiation by ozone with the cooling due to the emission of infrared radiation by carbon dioxide. With a simple calculation we then deduced a temperature at about 50 km altitude of 280 K, approximately as found at that level. If a concentration of carbon dioxide of 560 ppm is inserted in equation (4.26) the cooling rate calculated in equation (4.27) increases and the temperature at 50 km altitude drops by about 20°C (problem 14.4). The effect on the stratosphere of increased atmospheric carbon dioxide is therefore a cooling.

Turning now to the effect on the troposphere which to a first approximation is decoupled from the stratospheric changes, we need first to estimate the change in outgoing radiation from carbon dioxide at the top of the troposphere. Reference again to fig. 12.7 shows that it is radiation from the wings of the 15 μm carbon dioxide band that originates from the troposphere. The distribution in height of the origin of the radiation leaving the top of the atmosphere in narrow spectral intervals of the carbon dioxide band will be similar to those illustrated for the microwave oxygen band in fig. 12.10(b). If the concentration of carbon dioxide doubles, the outgoing radiation from the wings of the carbon dioxide band will originate from higher levels. Because of the fall of temperature with height these higher levels are cooler and therefore the amount of emitted radiation falls.

The size of this fall in emitted radiation can be estimated as follows. Problem 12.4 works out for a frequency in the wing of a collision broadened line the form of the contribution function (as illustrated by the curves in fig. 12.6) which describes the distribution with height of the origin of the radiation leaving the atmosphere in a given spectral interval. The pressure p_m at the level where the contribution is a maximum is given by

$$p_m = \left\{ \frac{2g\pi p_0 (\nu - \nu_0)^2}{s\gamma_0 c} \right\}^{1/2} \tag{14.1}$$

Note that p_m is proportional to $c^{-0.5}$. This will also be true when the radiation is added up from a number of frequencies (cf. problem 12.5). If therefore the concentration c is doubled, the origin of the radiation moves up to a level where the pressure is less by a factor of $2^{-0.5}$ – about 3 km higher – where, assuming a lapse rate of 6°C/km, the temperature on average will be about 18°C less. If we assume an average radiating temperature of 250 K for the undisturbed case, it will be about 232 K in the doubled CO_2 case. With this temperature difference the drop in the radiation emitted within the interval 700–750 cm^{-1} (the approximate spectral region involved) will be about 3 W m^{-2}.

This, of course, is a very crude estimate. A proper calculation will take account of the detailed line structure of the carbon dioxide band

Fig. 14.10. Illustrating the enhanced greenhouse effect. Under natural conditions (a) the net solar radiation coming in (S = 235 W m^{-2}) is balanced by thermal radiation (L) leaving the top of the atmosphere; average surface temperature (T$_s$) is 15°C. If the carbon dioxide concentration is suddenly doubled (b), L is decreased by 4 W m^{-2}. Balance is restored if nothing else changes (c) apart from the temperature of the surface and lower atmosphere which rises by 1.2°C. If feedbacks are also taken into account (d) the average temperature of the surface rises by about 2.5°C.

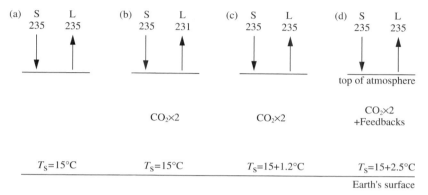

including the other similar spectral region on the long-wave side of the band that is partially obscured by water vapour lines. It will also allow for the detailed vertical structure of the atmosphere and the average cloud cover. Further, since what is required is the change in net radiation at the top of the troposphere, allowance has also to be made for the change in the downward radiation from the stratosphere resulting from the changes there. When all these are properly taken into account, it is found that the best estimate for the change on average in the net radiation at the top of the troposphere due to a doubling of the carbon dioxide concentration – assuming everything else remains the same – is about 4 W m^{-2}.

This change in radiation at the top of the troposphere is known as *radiative forcing*. The effect of different greenhouse gases can conveniently be compared by comparing their contributions to the radiative forcing.

Let us continue with this thought experiment of a doubling of the carbon dioxide concentration. The imbalance of 4 W m^{-2} at the top of the troposphere creates an imbalance in the energy budget there. To restore balance the temperature of the lower atmosphere and the surface will rise. If we assume that this occurs with no change in lapse rate, cloudiness or water vapour, the average temperature of the surface and the lower atmosphere will rise by approximately 1.2°C (problem 14.5). Fig. 14.10 presents this result.

Note that the net change of radiation at the top of the troposphere is approximately logarithmic with respect to the carbon dioxide concentration (problem 14.10). A quadrupling of carbon dioxide concentration for example will lead to an imbalance of $8\,W\,m^{-2}$. Removal of carbon dioxide altogether leads to a temperature fall of about 6°C (problem 14.7).

We find therefore that the influence of increased carbon dioxide on atmospheric temperature is to cool the stratosphere and mesosphere and to warm the lower atmosphere and the surface. An important factor in this warming of the troposphere is the existence of a falling lapse rate of temperature with height which, of course, results from the influence of the natural greenhouse effect (§2.5).

The calculation so far in this section provides a basic description of the enhanced greenhouse effect. However, it has ignored any consequential changes that may occur in other parts of the atmosphere or the climate system. In reality many other factors will change, some of them in ways that add to the warming (these are positive feedbacks) and some that reduce the warming (negative feedbacks). The situation is therefore much more complicated than this simple calculation. The following paragraphs will be addressing the nature and magnitude of some of the feedbacks. Suffice it to say here that, when these feedbacks are allowed for, the best estimate at the present time (see IPCC, 1996 and IPCC, 2001) of the increased average temperature of the earth's surface if the carbon dioxide concentration were to be doubled is about twice that of the simple calculation – or 2.5°C (with a range of uncertainty of 1.5–4.5°C). Comparing this change in global average temperature with changes which have typically occurred in the past (for instance that of 5 or 6°C between the middle of an ice age and the warm periods in between) we see that it is a large change. Further, if this change in global average temperature were to occur in less than a century, it would be a very rapid change compared to most rates of change in the past.

Fig. 14.11 compares the radiative forcing from carbon dioxide and other greenhouse gases with that from likely changes in aerosols and in solar radiation since pre-industrial times. Note that, because ozone is a greenhouse gas, depletion of stratospheric ozone and increase in tropospheric ozone have led to radiative forcing. The direct radiative forcing due to the increase in aerosols from anthropogenic sources is mainly negative. A further way by which particles in the atmosphere can influence radiative forcing is through their effect on cloud formation which is described as *indirect* radiative forcing and which arises as follows. If particles are present in large numbers when clouds are forming, the resultant cloud will

Fig. 14.11. Estimates of the globally and annually averaged radiative forcings (in W m⁻²) due to a number of anthropogenic and natural factors from pre-industrial times (1750) to 2000. The height of the rectangular bars indicates mid-range estimates of the forcings; the error bars show estimates of the uncertainty ranges. Note that only the range of values, not a central estimate, is shown for the indirect aerosol forcing, because of its large uncertainty. The contributions of individual gases to the direct greenhouse forcing are indicated in the first bar; the forcings associated with changes in ozone are shown in the second and third bars. Estimates of the forcings (globally averaged values) due to changes in aerosols are shown in the following bars. The last two bars are estimates of the forcings associated with changes in land use because of the changes in albedo and with possible variations in the input of solar radiation. Also included is an index of scientific understanding for each forcing representing a subjective judgement about the degree of knowledge and understanding regarding the mechanisms involved. (From 'Summary for Policymakers', IPCC, 2001)

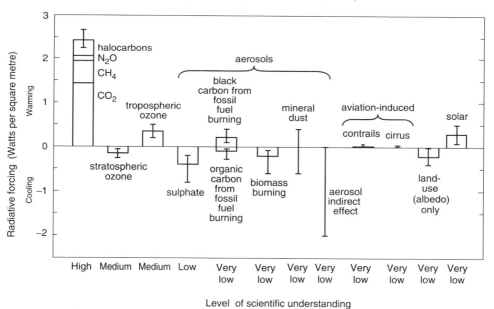

contain a large number of small drops – smaller than would otherwise be the case. This is similar to what happens as polluted fogs form in cities. Such a cloud is more highly reflecting to sunlight than one containing a smaller number of large particles, thus further increasing the energy loss resulting from the presence of the particles. The magnitude of this effect as it occurs in the atmosphere is difficult to estimate.

Note that radiative forcing due to changes in solar radiation or to changes in the well-mixed greenhouse gases is applied to the atmosphere on a global scale. In contrast the radiative forcing due to increases in aerosols or greenhouse gases such as tropospheric ozone that are variable in their distribution varies greatly on the regional scale. In these cases, the global average values of radiative forcing provide indications of their individual magnitudes, but they cannot be simply added or subtracted to obtain their total effect on the climate.

14.7 Feedback processes

The four main physical feedback processes we need to consider are due to changes in:

- atmospheric water vapour content – increased temperature leads to increased evaporation and hence on average increased water vapour amounts in the atmosphere (problem 14.8). Since water vapour is a strong greenhouse gas this is a positive feedback;
- atmospheric temperature structure particularly arising from changes in water vapour content – this is sometimes known as *lapse rate feedback* (problem 14.9). The average effect of this feedback on surface temperature tends to be negative;
- the cover of sea or land ice – if a part of the sea surface with an albedo of about 0.1 or a part of the land surface with an albedo of 0.3 is covered with ice or snow with an albedo of 0.6 or more, the amount of solar energy absorbed at the surface may be more than halved causing the surface to cool further; this is a positive feedback known as ice-albedo feedback;
- cloud cover – since clouds have a strong effect on the absorption, reflection or emission of radiation this is often known as cloud radiation feedback; it is described in more detail in §6.5. The sensitivity of climate to possible changes in cloud amount or structure is illustrated in table 14.1 where the hypothetical effect on the climate of changes of a few per cent in cloud cover is compared with the expected changes due to a doubling of the carbon dioxide concentration.

Because the lapse rate feedback arises mostly through changes in water vapour content (problem 14.9), the first two of these feedback processes are sometimes lumped together under the water vapour feedback heading.

In addition to these physical feedbacks there are subsequent changes in the atmospheric and ocean circulations, or dynamical feedbacks, which will alter the pattern of change over the earth's surface (see §§14.8 and 14.9). Further there are biochemical feedbacks arising from changes in the biosphere (e.g. in vegetation) which can lead in

Table 14.1. *Estimates of global average temperature changes under different assumptions about changes in greenhouse gases and clouds.*

Greenhouse gases	Clouds	Change in °C from current average global surface temperature of 15°C
As now	As now	0
None	None	−21
As now	None	4
As now	As now but +3% high cloud	0.3
As now	As now but +3% low cloud	−1.0
Doubled CO_2 concentration; otherwise as now	As now (no additional cloud feedback)	1.2
Doubled CO_2 concentration + best estimate of feedbacks	Cloud feedback included	2.5

turn to changes in atmospheric composition. These will not be discussed further here; for further information refer to the references given in the bibliography.

It is helpful to compare different feedback processes by employing a feedback parameter δ defined by

$$\delta = \frac{\Delta(F_L - F_S)}{\Delta T_s} \tag{14.2}$$

where F_L and F_S are the long-wave and solar radiation components of the net radiation at the top of the troposphere and T_s is the surface temperature. Using the values we quoted above for the change in T_s of 1.2°C for a change in $F_L - F_S$ of 4 W m^{-2} the value of δ resulting from the basic temperature response to the net radiation change (this is sometimes called the temperature feedback) is 3.3 W m^{-2} K^{-1}. From experiments with numerical models of the climate (see §14.8) in which the values of sea surface temperature have been deliberately perturbed (e.g. by +2°C and −2°C) it is possible to estimate the contributions to ΔF_L and ΔF_S, and hence to the feedback parameter δ which arise from some of the different feedback processes mentioned above. One particular study (Zhang et al., 1994) provides values of these contributions averaged over the globe of about −1.6 W m^{-2} K^{-1} for the water vapour feedback parameter and about 0.4 W m^{-2} K^{-1} for the lapse rate feedback parameter. Although the individual values of these two parameters depend on the assumptions made in any particular model, their sum is substantially model independent.

In §6.5 were described the two effects of clouds on atmospheric radiation fluxes. On the one hand, they reflect solar radiation and reduce the amount of solar energy reaching the surface, on the other hand they strongly absorb infrared radiation and so decrease the net loss of long-wave radiation by the surface. Chapter 6 also introduced the concept of Cloud Radiative Forcing (*CRF*). It is helpful to describe model results as they relate to cloud radiation feedback in terms of *CRF*.

From the model experiments of the kind described above the difference in *CRF*, $\Delta(CRF)$, between the cases with $\Delta T_s = 2°C$ and $-2°C$ averaged over the globe can be derived. Typical current models possess values of $\Delta(CRF)/G$ (where G is the overall change in net radiation at the top of the troposphere between the two sea surface temperature cases) between about 0.5 and −0.2. Climate models therefore can show an average cloud feedback which may be either positive or negative; further the sign of the feedback shows substantial regional variation. Uncertainty regarding cloud radiation feedback is the main reason for the wide range of 1.5°C to 4.5°C in the estimate for the change in global average surface temperature due to a doubling of carbon dioxide concentration.

14.8 Modelling climate change

So far we have been considering the influence of the enhanced greenhouse effect on a vertical column of atmosphere representing a global average – a very simple one dimensional model. This has provided a crude but useful indication of the overall response so far as temperature change is concerned. A relatively simple extension of the one dimensional model is to allow for variations with latitude and set up a two dimensional model. But the climate response to the enhanced greenhouse effect is far from uniform over the globe. For a more detailed and accurate description of climate change we need to employ models which are much more elaborate and which adequately include descriptions of structure and dynamics in all three dimensions.

In chapter 11 we described numerical models of the atmosphere and its circulation. Models of the climate are constructed in the same way except that they need to include the processes involved in changes which occur in all the components and interactions illustrated in fig. 14.1. Also, in order that they can be run over much longer periods (for instance, for the study of anthropogenic climate change, for several centuries), for a given computer power, the spatial resolution needs to be substantially

Fig. 14.12. Component elements and parameters of a coupled atmosphere–ocean model including exchange at the atmosphere–ocean interface.

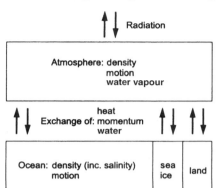

degraded. At the time of writing a computer climate model running on the largest computer available (fig. 11.4) possesses a typical resolution of 250 km in the horizontal and about 30 levels in the vertical.

We have already indicated the importance of adequately describing in models physical processes in the atmosphere, for instance those associated with clouds. Of particular importance in climate models is that there be adequate description of the coupling of atmosphere and ocean processes and dynamics at the atmosphere–ocean interface. Early climate models only included the ocean very crudely by representing it as a simple slab some 50 or 100 metres deep, the approximate depth of the 'mixed layer' of ocean which responds to the seasonal heating and cooling at the earth's surface. In such models the transport of heat by ocean currents could not be explicitly described; adjustments were made to allow crudely for it.

More recent models couple the circulations of the atmosphere and the ocean; fig. 14.12 shows the ingredients of such a model. At the ocean–atmosphere interface there is exchange of heat, water and momentum. We mentioned in §14.3 that evaporation of water from the ocean surface into the atmosphere is an important energy source for the atmosphere's circulation and we pointed out the strong influence of the patterns of sea surface temperature on regional climate.

The ocean structure and circulation is, for its part, strongly determined by atmospheric conditions at the surface. Momentum introduced from the atmosphere's motion near the surface and the net heat input from the atmosphere to the ocean are important drivers for the ocean's circulation. The introduction of fresh water into the ocean through precipitation from the atmosphere also influences the ocean's circulation (and hence the climate) through its effect on the distribution of salt, which in turn affects the ocean density. A dramatic example of this occurs in the north Atlantic Ocean between Iceland and Scandanavia where because the water is unusually cold and salty and therefore of higher density it tends to sink, providing an important source of water for a global scale slow circulation in the deep ocean. This circulation is known as the thermo-haline circulation. Evidence from paleoclimate history shows that, at times in the past, because of increases in the input of fresh water into the north Atlantic, for instance from unusual melting of ice, this circulation has been weakened or even cut off. Climate models indicate that the increased precipitation expected in the warmer world of increased greenhouse gases could be sufficient to act in the same way and substantially weaken the thermo-haline circulation. This would in turn tend to alter the amount of heat transported northward by the Atlantic Ocean with significant consequences for the distribution of climate change especially in north-west Europe.

Once a comprehensive climate model has been formulated it can be tested in three main ways. Firstly, it can be run for a number of years of simulated time and the climate generated by the model compared in detail to the current climate, both in terms of averaged quantities and variability (fig. 14.2(a)). Secondly, it can be used to predict the effect of large perturbations on the climate, for instance volcanic eruptions (fig. 14.13). Good progress is also for instance being achieved with the prediction of El Niño events and the associated climate anomalies up to a year ahead. Thirdly, model simulations can be compared against observations for past climates, for instance for periods when the configuration of the earth's orbit was substantially different from the present (cf. figs. 14.3 and 14.4). Climate models which are currently employed for climate prediction stand up well to such comparisons.

The largest interest in climate models is in their ability to predict the likely climate change resulting from the increase in greenhouse gases due to human activities. They do this by comparing with details of the current climate the detailed climate simulated by a model in which the concentrations of greenhouse gases (especially carbon dioxide) are increased according to some projected future scenario of the extent of the human activities concerned, for instance, with fossil fuel use and deforestation. Examples of such simulations of the changes in global average surface temperature for a range of projections of greenhouse gas emissions for the twenty-first century are shown in fig. 14.14.

Fig. 14.13. The predicted and observed changes in global surface air temperature after the eruption of Mount Pinatubo, expressed as three-monthly averages from April–June 1991 to April–June 1995. (After Hansen et al., 1992)

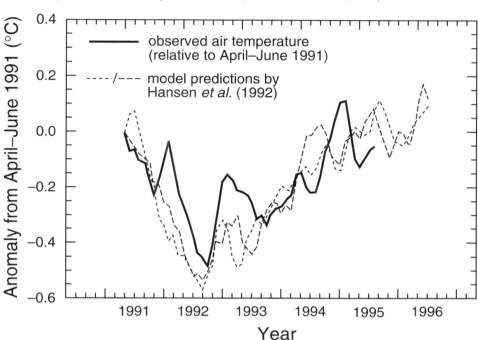

14.9 Observations of climate

In order to develop climate models, observations are required of all parts of the climate system (fig. 14.1) and of the processes that occur within them. Further, as we noted in the previous section observations are necessary to validate climate models. Chapter 12 provided details of global observations of the atmosphere and the earth's surface from both surface-based and space-based instruments. It is also required that the ocean, the cryosphere and the land and ocean biosphere are adequately observed. The Global Climate Observing System (GCOS) has been set up internationally to assist in the definition, coordination and implementation of the necessary observations.

Observations of the ocean are particularly important and are coordinated through the Global Ocean Observing System (GOOS). Observations of the surface, for instance of temperature and wind, especially from space vehicles were mentioned in chapter 12. Other measurements of the ocean from space of importance for climate are of ocean colour (relevant to ocean biology) and of ocean surface topography. Changes in the latter,

Fig. 14.14. (a) Atmospheric concentrations of CO_2 from 1000 to 2000 as measured from ice core data and, for the recent period, direct sampling. Estimates of concentrations to 2100 from a carbon cycle model with input from a number of projections of CO_2 emissions from fossil fuel and other sources (more detail of these CO_2 emission projections in IPCC, 2001). (b) Variations of the earth's global average surface temperature from 1000 to 2000 as in fig. 14.2. Estimates of temperature to 2100 based on forcing from the CO_2 concentrations in (a) applied to a climate model with average sensitivity to CO_2 increase. The envelope shows the range of temperature estimates covering the various projections of CO_2 emissions in (a) and the results from a number of models with a range of sensitivity to CO_2 increase. The bars on the right show, for the six scenarios, the range in 2100 produced by a number of climate models having a range of sensitivity to CO_2 increase. The grey envelope shows the complete range including the different scenarios and the different models.

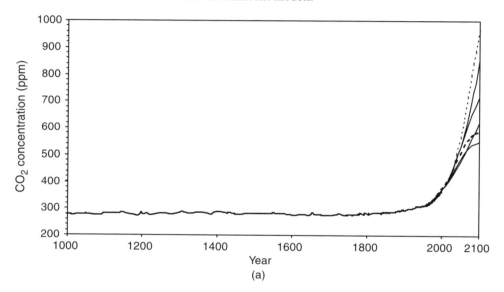

(a)

which can be measured with a precision of a few cm with radar altimetry, are related to ocean currents in a similar way to that in which changes in atmospheric pressure are related to atmospheric winds. Below the ocean surface, measurements of density (which involves salinity as well as temperature), composition and dynamics are made with instruments deployed from ships, buoys or submersibles.

14.10 Dynamical response to external forcing

The discussion concerning the greenhouse effect earlier in the chapter that calculated the effect on the surface temperature of changes in net radiation input at the top of the atmosphere was a one dimensional

Fig. 14.14. (*continued*)

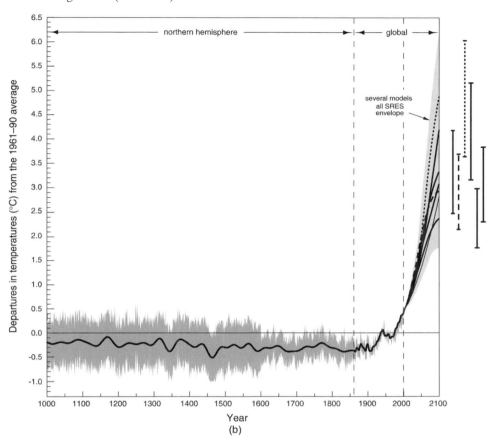

Year
(b)

discussion and ignored changes in atmospheric circulation. A diagram such as that in fig. 14.14(b), which results from averages over the globe obtained from a fully three dimensional global circulation model, demonstrates that to a substantial degree the results of the one dimensional calculation are borne out in the global average.

However, as the model results readily demonstrate, the detailed pattern over the globe of climate response to changes in net radiation at the top of the atmosphere, in terms for instance of surface temperature, is anything but uniform. This is not surprising when it is realized that the response of the atmospheric dynamical system to particular forcing is strongly non-linear. Take for instance the very simple model of atmospheric circulation set up by Lorenz and described in §13.2. In fig. 14.15 are shown examples of the Lorenz attractor perturbed by a weak external

Fig. 14.15. Showing cross-sections of the Lorenz attractor in the X-Y plane under conditions of external forcing acting in two different directions in phase space shown by the arrows (after Palmer, 1999). When unforced the two lobes of the attractor are approximately equal.

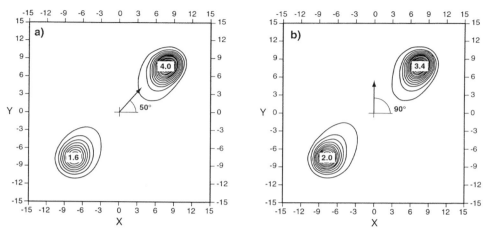

forcing term applied in the x-y plane in different directions. In the absence of external forcing, the probability density function (PDF, which describes the likelihood of any given 'circulation') in the neighbourhood of the two density maxima (corresponding to the two 'butterfly wing' regimes of the Lorenz attractor) is identical in character. The major component of response to the forcing whatever its direction is associated with an increase of the PDF in one regime and a decrease of the PDF in the other regime. The locations or shapes of the regimes do not alter. It can be argued (Palmer, 1999) that similarly much of the detailed atmospheric response to external forcing (e.g. by solar variations, volcanic eruptions or by changes in the concentrations of greenhouse gases) will be manifest in changes in the intensity and frequency of occurrence of some of the preferred naturally occurring circulation regimes which are similarly non-linear in character.

Because of its large influence on manifestations of extreme climate, the El Niño is the most important of these regimes. Of especial interest is the possibility of changes in the intensity, frequency or character of the El Niño due to increased forcing from greenhouse gases. Such possibilities are suggested by the strong El Niño events experienced during the 1990s (fig. 14.6(a)) and are currently being explored by model simulations.

14.11 The impacts of climate change

The main impacts of climate change due to a substantial increase in greenhouse gases are expected to be the following.

(1) The projected rate of global average surface temperature rise during the twenty-first century is very likely to be greater than any that has occurred during at least the last 10 000 years. Nearly all land areas will warm more rapidly than the global average. There will be an increase in the number of very hot days and hot periods especially in major continental areas.

(2) A rise in sea level will occur due to expansion of the oceans with increasing temperature and because of melting of glaciers. It is estimated to be about 0.5 m on average for doubled carbon dioxide concentration. The main ice sheets of Antarctica and Greenland are not likely to change much in overall size; ice added because of increased precipitation will tend to be balanced by ablation of ice around the margins.

(3) Warmer temperatures will lead to increased evaporation, greater average water vapour concentrations and an increase on average in overall precipitation (estimated to be of the order of 5% in global average for doubled carbon dioxide concentration).

(4) The increased latent heat release on condensation will lead on average to a more intense hydrological cycle. This will mean an increased likelihood of heavy rainfall (and hence floods) in some places and, because of drier air in regions of downdraught, to an increased likelihood of reduced rainfall (and hence periods of drought) in other regions. Substantial changes in water resources, water availability and the distribution of soil moisture are therefore expected.

The effect of these impacts on human activities is expected to be substantial. In particular many humans and ecosystems will be unable to adapt easily to the expected rate of change. To mitigate the changes, emissions of greenhouse gases, especially carbon dioxide, need to be substantially reduced. Appropriate action is being addressed both internationally and nationally. Further details of the science, the impacts and the possible policy responses can be found in Houghton (1997), IPCC (2001), and on the IPCC web pages, www.ipcc.ch.

Problems

14.1 In §14.3 it was shown that the intensity of solar radiation incident
on the earth and its distribution over the earth varies due to secular
changes in the earth's orbit about the sun.

(a) Show from a consideration of the geometry of the earth's
orbit that the fractional variation in the incidence of solar
radiation at the solstices from its mean value, due to variations in distance from the Sun, is approximately $2e \sin \omega$.
From fig. 14.4 what are the percentage differences in the
intensity of solar radiation from its current value at the
summer and winter solstices 12000 years B.P. and 120000
years B.P.

(b) The zenith angle z of the sun at any given time of day is
given by

$$\cos z = \sin \phi \sin \delta + \cos \phi \cos \delta \cos h \qquad (14.3)$$

where ϕ is the latitude, δ the solar declination and h the hour angle.
Integrate the solar radiation incident over a day on a horizontal
surface at the top of the atmosphere to find the percentage change
in daily insolation at latitude $50°$ at the summer and winter solstices
from its current value at 12000 B.P. and at 12000 B.P. due to changes
in the obliquity factor (see fig. 14.4).

(c) Remembering that for the orbit of a planet around the
sun, equal areas are swept out in equal times, show that the
difference between the length of the summer half-year t_s
and the winter half-year t_w is given by

$$t_s - t_w = \frac{4}{\pi} t e \sin \omega \qquad (14.4)$$

where t is the length of the year.

14.2 Equation (1.1) for the radiative balance of the earth as a whole can
be written in the form

$$\frac{1}{4} F(1 - A) = \varepsilon \sigma T_s^4 \qquad (14.5)$$

where F is the solar constant, A the albedo, T_s an average surface
temperature ($\sim$280 K) and ε a quantity allowing for the effect of
clouds and gaseous components contributing to the greenhouse
effect (§2.5).

From equation (14.5) show that for constant A and ε, i.e. for no cloud radiation feedback, a 1% change in F leads to a 0.7 K change in T_s. Show also that a 0.01 change in A leads to a 1 K change in T_s. Assuming that the mixed layer of the oceans with a mean depth of ~100 m is involved in these changes in addition to the atmosphere, what is the time constant for the change in T_s resulting from the changes in F or in A?

14.3 Between 18 000 B.P., the time of the last glacial maximum, and 6000 B.P. it is estimated that $40 \times 10^6 \, \text{km}^3$ of ice melted from the ice sheets covering the continents. What percentage increase in solar radiation would be required polewards of latitude 45° to provide the heat required to melt the ice during the period assuming all the increase were available for this purpose?

14.4 Insert a CO_2 mixing ratio of double the pre-industrial value (560 ppmv, $c = 7.8 \times 10^{-4}$) in equation (4.26), and calculate the new cooling rate in (4.27) and, assuming the ozone heating is unchanged, the new equilibrium temperature at 50 km.

14.5 The average net infrared radiation leaving the top of the troposphere is approximately 235 W m^{-2}. Doubling the carbon dioxide concentration reduces this by 4 W m^{-2}. The radiation originating at all levels in the troposphere will rise to restore the balance. Estimate the temperature increase that will restore this balance. In making the estimate you will wish to use an average radiating temperature for the earth. Table 1.2 suggests this should be 250 K. This value averages over cloudy and clear regions. For clear regions the average radiating temperature is about 270 K. It is this latter value which is more appropriate to use in this calculation. Explain the reason for this.

14.6 The chlorofluorocarbons (CFCs) possess absorption bands near 11 μm in the infrared atmospheric window. Increases in the concentrations of these compounds in the atmosphere add to the greenhouse effect. In the absence of feedback effects, estimate the change in the net radiation at the top of the atmosphere which would occur if the combined concentrations of the CFCs in the troposphere were 1 part in 10^9, given that the strength of the bands involved near 11 μm is ~2000 cm^{-1}(atm-cm)$^{-1}$. Hence estimate the change in surface temperature which might occur.

14.7 Estimate the effect on the surface temperature of completely removing the carbon dioxide from the atmosphere. You can do this by referring to fig. 12.7(b) and working out the difference in the net radiation at the top of the troposphere if the $15\,\mu m$ CO_2 band is removed.

14.8 Estimate a value for the water vapour feedback parameter as follows. For a temperature rise of $1°C$ the saturated water vapour pressure rises by about 7%. Suppose therefore for a temperature rise of $1°C$ the water vapour content of the atmosphere rises in the same proportion. Assume emission from water vapour dominates at all wavenumbers below $600\,cm^{-1}$ (wavelengths beyond $16.7\,\mu m$) (see fig. 12.7). Then applying equation (14.1), in a similar way as was done in the text for CO_2, show that the resulting change in net radiation at the top of the troposphere is about $1.8\,W\,m^{-2}$ in reasonable agreement with the value found in model studies quoted in the text.

14.9 The lapse rate in tropical regions is closely linked to the saturated adiabatic which reduces with increasing temperature. What effect will this have on the calculation of the increase in surface temperature if the atmospheric CO_2 concentration is doubled? Is lapse rate feedback under these conditions positive or negative?

14.10 Show that for changed atmospheric carbon dioxide concentration, the resulting changes in net radiation at the top of the troposphere and in surface temperature are approximately logarithmic with respect to the carbon dioxide concentration (use equation (14.1), and the hydrostatic equation (1.4)).

14.11 Volcanic eruptions often inject substantial amounts of dust into the stratosphere. If the particles are of sufficiently small size these layers can persist for many months or years. Suppose such a layer reflects 1% of the incident solar radiation, while, because of the small size of the particles, having a negligible effect on the long-wave radiation emitted from below. What change in surface temperature might be expected resulting from this dust layer?

14.12 Measurements may be made from satellites of the distribution of the flux F_{TA} of net radiation at the top of the atmosphere. Show that the distribution of the flux F_{BA} of net radiation at the bottom of the atmosphere can be written as

$$F_{BA} = F_{TA} - \mathrm{div}\ T_A - S_A \qquad (14.6)$$

where T_A is the horizontal transport of energy by the atmosphere and S_A is the rate of storage of energy in the atmosphere.

If T_O and S_O are respectively the horizontal transport of energy and the rate of storage of energy in the oceans, show that, over the oceans,

$$\mathrm{div}\ T_O = F_{BA} - S_O \qquad (14.7)$$

Oort and Vonder Haar (1976) used these equations together with observations of F_{TA}, S_A and S_O to make an estimate of the transport of heat in the oceans averaged across circles of latitude and its variation with time of year. This is an important quantity in understanding the role of the oceans in climate.

14.13 Various attempts have been made to investigate the stability of the climate system by setting up simple global energy balance models that incorporate parametrizations based on empirical information.

Making use of empirical relations proposed by Budyko (1969) and Sellers (1969) (see also Wiin-Nielsen, 1981), set up a simple global averaged model as follows. Take the incoming radiant energy F_{in} as given by

$$F_{in} = F_s(1 - A(T))$$

where F_s is the solar radiation on a surface normal to the solar beam outside the atmosphere and A is the average planetary albedo. A is considered to be influenced mainly by the snow and ice cover and to be a function only of the average surface temperature T, following the empirical relations:

$$A(T) = A_{max} = 0.85 \quad T < T_1 = 216\mathrm{K}$$
$$A(T) = A_{max} - \frac{A_{max} - A_{min}}{T_2 - T_1}(T - T_1) \quad T_1 < T < T_2$$
$$A(T) = A_{min} = 0.25 \quad T \geq T_2 = 283\mathrm{K}$$

Take the outgoing infrared radiation to space as

$$F_{out} = \sigma T^4\left[1 - \frac{1}{2}\tanh 19 \times 10^{-16} T^6\right]$$

where the term in brackets represents in a crude way the effect of water vapour, carbon dioxide and clouds on terrestrial radiation.

Plot as a function of surface temperature T the function $F_{in} - F_{out}$. Show that there are three equilibrium states corresponding to a completely ice-covered earth, to partial ice cover and to an ice-free earth. Show also that two of the equilibrium positions are stable with respect to small changes while the other is unstable.

Now reduce the solar constant F_s by 12% and show that only one equilibrium state remains.

Appendices

1. Some useful physical constants and data on dry air

Avogadro's number	$6.022 \times 10^{26}\,\text{kmol}^{-1}$
Loschmidt number	$2.687 \times 10^{25}\,\text{m}^{-3}$
Boltzmann constant k	$1.381 \times 10^{-23}\,\text{J}\,\text{K}^{-1}$
Planck constant h	$6.6262 \times 10^{-34}\,\text{J}\,\text{s}$
Gas constant R	$8.3143\,\text{kJ}\,\text{K}^{-1}\,\text{kmol}^{-1}$
Stefan–Boltzmann constant σ	$5.670 \times 10^{-8}\,\text{J}\,\text{m}^{-2}\,\text{K}^{-4}\,\text{s}^{-1}$
Velocity of light c	$2.998 \times 10^{8}\,\text{m}\,\text{s}^{-1}$
Ice point	$273.15\,\text{K}$
Earth's mean radius	$6371\,\text{km}$
Mean solar angular diameter	31.99 minutes of arc
Standard surface gravity	$9.806\,65\,\text{m}\,\text{s}^{-2}$
Standard pressure p_0	$1.013\,25 \times 10^{5}\,\text{Pa}\ (\equiv 1013.25\,\text{mb})$

Data on dry air

Apparent molecular weight	28.964
Gas constant for dry air	$287.05\,\text{J}\,\text{kg}^{-1}\,\text{K}^{-1}$
Specific heats of dry air:	
at constant pressure c_p	$1005\,\text{J}\,\text{kg}^{-1}\,\text{K}^{-1}$
at constant volume c_v	$718\,\text{J}\,\text{kg}^{-1}\,\text{K}^{-1}$
Ratio of specific heats γ	1.40
Density of dry air at $273\,\text{K}$ and $101.3\,\text{kPa}$ pressure	$1.293\,\text{kg}\,\text{m}^{-3}$
Viscosity (at STP)	$1.73 \times 10^{-5}\,\text{kg}\,\text{m}^{-1}\,\text{s}^{-1}$
Kinematic viscosity (at STP)	$1.34 \times 10^{-5}\,\text{m}^{2}\,\text{s}^{-1}$
Thermal conductivity (at STP)	$2.40 \times 10^{-2}\,\text{W}\,\text{m}^{-1}\,\text{K}^{-1}$

Refractive index n of dry air at 101.3 kPa, 273 K, and wavelength of $1\,\mu$m $= 1 + 289.2 \times 10^{-6}$

At other wavelengths $\lambda\,\mu$m Edlén's formula (applying to a temperature of 288 K) may be used

$$\{n(\lambda) - 1\} \times 10^6 = 64.328 + 29498.1(146 - \lambda^{-2})^{-1} + 255.4(41 - \lambda^{-2})^{-1}$$

To obtain n at other temperatures and pressures note that $n - 1$ is proportional to density. For air containing water vapour there is a correction which is usually negligible at visible and infrared wavelengths but which becomes significant in the mm wavelength region. (For more information see Penndorf, 1957.)

2.   Properties of water vapour

Molecular weight	18.015
Latent heat of fusion at 273 K	$3.34 \times 10^5\,\mathrm{J\,kg^{-1}}$
Latent heat of vaporization at 273 K	$2.500 \times 10^6\,\mathrm{J\,kg^{-1}}$
Specific heat of liquid water at 273 K	$4.218 \times 10^3\,\mathrm{J\,kg^{-1}\,K^{-1}}$
Specific heat of ice at 273 K	$2.106 \times 10^3\,\mathrm{J\,kg^{-1}\,K^{-1}}$
Density of ice at 273 K	$917\,\mathrm{kg\,m^{-3}}$

Table A2. After *Smithsonian Meteorological Tables*, Smithsonian Institute, Washington, D.C. 1958.
Saturation vapour pressure (Pa) *over pure liquid water*

°C	−30	−20	−10	0	+10	+20	+30	+40
0	50.88	125.40	286.27	610.78	1227.2	2337.3	4243.0	7377.7
+1	55.89	136.64	309.71	656.62	1311.9	2486.1	4492.7	7780.2
+2	61.34	148.77	334.84	705.47	1401.7	2643.0	4755.1	8201.5
+3	67.27	161.86	361.77	757.53	1496.9	2808.6	5030.7	8642.3
+4	73.71	175.97	390.61	812.94	1597.7	2983.1	5320.0	9103.4
+5	80.70	191.18	421.48	871.92	1704.4	3167.1	5623.6	9585.5
+6	88.27	207.55	454.51	934.65	1817.3	3360.8	5942.2	10089
+7	96.49	225.15	489.81	1001.3	1936.7	3564.9	6276.2	10616
+8	105.38	244.09	527.53	1072.2	2063.0	3779.6	6626.4	11166
+9	115.00	264.43	567.80	1147.4	2196.4	4005.5	6993.4	11740

Saturation vapour pressure (Pa) *over pure ice*

°C	−100	−90	−80	−70	−60
0	1.403×10^{-3}	9.672×10^{-3}	5.472×10^{-2}	2.615×10^{-1}	1.080
+1	1.719×10^{-3}	1.160×10^{-2}	6.444×10^{-2}	3.032×10^{-1}	1.236
+2	2.101×10^{-3}	1.388×10^{-2}	7.577×10^{-2}	3.511×10^{-1}	1.413
+3	2.561×10^{-3}	1.658×10^{-2}	8.894×10^{-2}	4.060×10^{-1}	1.612
+4	3.117×10^{-3}	1.977×10^{-2}	1.042×10^{-1}	4.688×10^{-1}	1.838
+5	3.784×10^{-3}	2.353×10^{-2}	1.220×10^{-1}	5.406×10^{-1}	2.092
+6	4.584×10^{-3}	2.796×10^{-2}	1.425×10^{-1}	6.225×10^{-1}	2.380
+7	5.542×10^{-3}	3.316×10^{-2}	1.662×10^{-1}	7.159×10^{-1}	2.703
+8	6.685×10^{-3}	3.925×10^{-2}	1.936×10^{-1}	8.223×10^{-1}	3.067
+9	8.049×10^{-3}	4.638×10^{-2}	2.252×10^{-1}	9.432×10^{-1}	3.476

°C	−50	−40	−30	−20	−10	0
0	3.935	12.83	37.98	103.2	259.7	610.7
+1	4.449	14.36	42.13	113.5	283.7	
+2	5.026	16.06	46.69	124.8	309.7	
+3	5.671	17.94	51.70	137.1	337.9	
+4	6.393	20.02	57.20	150.6	368.5	
+5	7.198	22.33	63.23	165.2	401.5	
+6	8.097	24.88	69.85	181.1	437.2	
+7	9.098	27.69	77.09	198.4	475.7	
+8	10.21	30.79	85.02	217.2	517.3	
+9	11.45	34.21	93.70	237.6	562.3	

3. Atmospheric composition

Table A3. *Normal composition of clean dry air near sea level.*

Gas	Volume mixing ratio
Nitrogen (N_2)	0.78083
Oxygen (O_2)	0.20947
Argon (Ar)	0.00934
Carbon dioxide (CO_2)	0.00037
Neon (Ne)	18.2×10^{-6}
Helium (He)	5.2×10^{-6}
Krypton (Kr)	1.1×10^{-6}
Xenon (Xe)	0.1×10^{-6}
Hydrogen (H_2)	0.5×10^{-6}
Methane (CH_4)	1.8×10^{-6}
Nitrous oxide (N_2O)	0.3×10^{-6}
Carbon monoxide (CO)	0.1×10^{-6}

For ozone (O_3) distribution see §5.6.

Water vapour (H₂O)

The above table is for dry air. Average water vapour distribution is shown in fig. A3.1. The water vapour mass mixing ratio of the stratosphere is much more uniform than in the troposphere and is in the range 2 to 5×10^{-6}.

Fig. A3.1. Contours of average water vapour mass mixing ratio in $g\,kg^{-1}$. (After Newell et al., 1972)

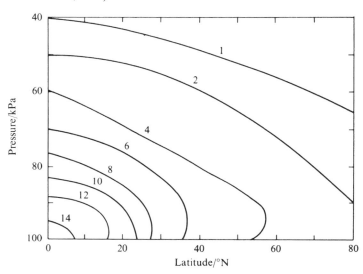

4.   Relation of geopotential to geometric height

Latitude	Geopotential metres (gpm)										
	10 000 m	20 000 m	30 000 m	40 000 m	50 000 m	100 000 m	200 000 m	300 000 m	400 000 m	500 000 m	600 000 m
0°	10 036	20 104	30 204	40 336	50 500	101 811	206 948	315 577	427 874	544 029	664 243
30°	10 023	20 077	30 163	40 282	50 432	101 672	206 656	315 115	427 225	543 174	663 161
45°	10 009	20 050	30 123	40 228	50 365	101 534	206 363	314 653	426 576	542 318	662 080
60°	9 996	20 024	30 083	40 174	50 297	101 395	206 071	314 191	425 927	541 465	661 000
90°	9 983	19 997	30 043	40 120	50 229	101 256	205 779	313 730	425 280	540 613	659 923

(From *American Institute of Physics Handbook*, McGraw-Hill 1972)

5.   Model atmospheres (0–105 km)

In the following tables geopotential height, pressure and temperature are plotted for different latitudes and months for typical northern hemisphere conditions. The three quantities are related by the hydrostatic equation (1.4).

The source of data is COSPAR (1990).

For an indication of the difference between the hemispheres see fig. 12.13.

The tabulated data of appendix 5 are plotted in fig. 5.1(*a*).

March

Pressure (kPa)	Geopotential height (km) 10 deg N	Temp (K)	Geopotential height (km) 40 deg N	Temp (K)	Geopotential height (km) 70 deg N	Temp (K)
100.0	0.096	299.9	0.130	283.9	0.118	256.4
95.0	0.544	297.3	0.551	280.9	0.504	256.1
90.0	1.013	294.7	0.992	277.9	0.908	255.6
85.0	1.504	291.9	1.454	274.8	1.333	254.8
70.0	3.141	283.0	2.991	265.7	2.762	250.0
50.0	5.862	267.7	5.537	250.5	5.160	236.2
40.0	7.585	256.8	7.140	238.9	6.674	226.5
30.0	9.689	241.7	9.095	225.9	8.542	218.2
20.0	12.43	219.5	11.72	217.3	11.12	217.7
10.0	16.56	195.7	16.09	214.9	15.55	217.5
7.0	18.61	200.9	18.34	214.8	17.83	217.0
5.0	20.60	207.5	20.45	215.3	19.97	217.1
3.0	23.75	216.3	23.69	217.5	23.24	218.4
1.0	30.97	233.1	30.82	227.7	30.32	223.0
0.7	33.45	241.9	33.24	235.4	32.67	227.2
0.5	35.88	250.1	35.60	243.1	34.93	231.9
0.3	39.70	261.6	39.32	254.5	38.46	239.8
0.1	48.37	271.9	47.78	266.8	46.51	259.6
0.7E-1	51.19	268.4	50.55	263.0	49.24	261.3
0.5E-1	53.81	262.6	53.11	256.5	51.80	259.3
0.3E-1	57.65	249.6	56.87	244.7	55.63	251.1
0.1E-1	65.28	226.3	64.46	229.8	63.37	231.7
0.7E-2	67.61	220.1	66.84	226.9	65.77	228.3
0.5E-2	69.76	215.5	69.07	224.3	68.01	226.0
0.3E-2	72.94	211.4	72.39	220.7	71.36	222.8
0.1E-2	79.71	208.6	79.35	209.8	78.44	214.4
0.7E-3	81.86	206.5	81.50	205.7	80.64	209.0
0.5E-3	83.89	202.2	83.51	200.8	82.67	201.2
0.3E-3	86.84	194.2	86.44	192.4	85.56	188.2
0.1E-3	92.93	186.6	92.40	180.9	91.30	174.5
0.7E-4	94.90	186.7	94.31	180.8	93.14	175.4
0.5E-4	96.77	186.8	96.12	182.0	94.90	178.5
0.3E-4	99.62	187.8	98.92	186.3	97.67	187.6
0.1E-4	106.09	205.2	105.51	215.0	104.51	230.3

June

Pressure (kPa)	Geopotential height (km) 10 deg N	Temp (K)	Geopotential height (km) 40 deg N	Temp (K)	Geopotential height (km) 70 deg N	Temp (K)
100.0	0.092	300.5	0.143	289.9	0.094	276.1
95.0	0.541	297.9	0.577	288.5	0.507	275.1
90.0	1.010	295.2	1.032	287.0	0.941	273.7
85.0	1.503	292.3	1.509	285.2	1.396	272.2
70.0	3.142	283.0	3.103	277.6	2.918	265.3
50.0	5.864	267.5	5.758	261.0	5.456	249.8
40.0	7.586	256.9	7.428	249.2	7.055	238.6
30.0	9.693	242.0	9.462	234.6	9.014	227.9
20.0	12.43	219.5	12.15	219.5	11.71	227.4
10.0	16.58	197.6	16.49	213.1	16.32	228.2
7.0	18.66	204.0	18.72	215.1	18.70	229.2
5.0	20.69	211.0	20.84	217.8	20.96	229.9
3.0	23.88	218.9	24.11	222.1	24.39	230.0
1.0	31.15	233.9	31.49	237.6	31.98	245.6
0.7	33.63	240.8	34.01	244.6	34.59	254.1
0.5	36.04	246.9	36.45	251.6	37.13	262.4
0.3	39.79	254.9	40.30	262.3	41.15	274.1
0.1	48.20	265.3	48.97	270.7	50.18	281.9
0.7E-1	50.97	263.2	51.77	266.1	53.10	277.9
0.5E-1	53.54	258.7	54.37	260.2	55.81	271.9
0.3E-1	57.34	249.1	58.18	249.0	59.80	260.1
0.1E-1	64.93	221.1	65.79	224.1	67.68	229.9
0.7E-2	67.19	213.0	68.09	216.2	70.02	220.0
0.5E-2	69.27	207.5	70.19	209.7	72.15	210.0
0.3E-2	72.34	203.8	73.26	201.0	75.17	194.2
0.1E-2	78.89	202.6	79.47	184.5	80.90	162.2
0.7E-3	80.98	201.2	81.35	180.3	82.52	153.5
0.5E-3	82.96	198.4	83.12	177.1	83.99	146.4
0.3E-3	85.88	192.5	85.73	172.9	86.09	140.5
0.1E-3	91.95	186.4	91.21	169.7	90.64	146.6
0.7E-4	93.93	186.4	93.01	172.4	92.22	153.2
0.5E-4	95.79	186.7	94.75	176.5	93.79	162.0
0.3E-4	98.65	188.6	97.51	186.2	96.38	181.1
0.1E-4	105.18	207.6	104.32	228.6	103.51	255.3

September

Pressure (kPa)	Geopotential height (km) 10 deg N	Temp (K)	Geopotential height (km) 40 deg N	Temp (K)	Geopotential height (km) 70 deg N	Temp (K)
100.0	0.088	300.5	0.146	293.2	0.094	275.3
95.0	0.536	297.7	0.583	290.9	0.505	273.5
90.0	1.006	294.9	1.040	288.4	0.935	271.6
85.0	1.497	292.0	1.519	285.8	1.386	269.5
70.0	3.134	282.7	3.119	277.2	2.892	262.1
50.0	5.854	267.5	5.777	261.5	5.402	247.4
40.0	7.577	256.9	7.452	249.8	6.984	236.2
30.0	9.683	241.7	9.493	235.3	8.920	225.0
20.0	12.42	219.3	12.18	219.7	11.57	223.4
10.0	16.58	198.5	16.50	211.2	16.10	224.1
7.0	18.67	204.6	18.71	214.0	18.43	223.5
5.0	20.70	210.8	20.82	217.1	20.62	222.8
3.0	23.87	216.3	24.07	220.1	23.94	221.8
1.0	31.08	232.9	31.31	231.6	31.14	228.6
0.7	33.55	239.6	33.75	236.9	33.55	232.4
0.5	35.95	246.6	36.12	243.1	35.87	236.8
0.3	39.72	257.4	39.83	253.4	39.47	245.1
0.1	48.25	268.4	48.24	265.6	47.67	262.0
0.7E-1	51.04	265.6	51.00	262.1	50.40	261.5
0.5E-1	53.63	260.1	53.55	255.2	52.96	257.4
0.3E-1	57.44	248.2	57.29	244.2	56.75	249.3
0.1E-1	65.05	226.1	64.84	225.9	64.52	232.9
0.7E-2	67.38	219.9	67.17	220.8	66.91	226.9
0.5E-2	69.53	215.3	69.33	216.8	69.13	221.3
0.3E-2	72.71	210.5	72.53	212.5	72.38	213.9
0.1E-2	79.41	206.5	79.28	204.9	79.06	200.7
0.7E-3	81.55	204.9	81.38	201.1	81.12	194.1
0.5E-3	83.56	201.1	83.35	195.8	82.99	185.2
0.3E-3	86.50	193.7	86.19	186.8	85.61	172.0
0.1E-3	92.58	186.5	92.00	177.1	90.93	164.5
0.7E-4	94.56	186.6	93.87	177.9	92.68	167.5
0.5E-4	96.42	186.8	95.65	180.0	94.37	172.6
0.3E-4	99.28	188.1	98.43	185.9	97.07	184.9
0.1E-4	105.77	207.0	105.10	220.0	103.98	237.3

December

Pressure (kPa)	Geopotential height (km) 10 deg N	Temp (K)	Geopotential height (km) 40 deg N	Temp (K)	Geopotential height (km) 70 deg N	Temp (K)
100.0	0.091	300.0	0.131	285.0	0.092	257.1
95.0	0.539	297.2	0.554	281.5	0.478	256.6
90.0	1.007	294.3	0.996	278.1	0.883	255.9
85.0	1.498	291.4	1.458	274.7	1.309	254.9
70.0	3.133	282.4	2.996	265.2	2.741	250.0
50.0	5.853	267.6	5.543	250.6	5.145	236.7
40.0	7.575	256.6	7.148	239.2	6.663	226.9
30.0	9.677	241.3	9.108	226.3	8.534	217.8
20.0	12.41	218.8	11.73	216.8	11.10	215.5
10.0	16.53	195.3	16.08	213.2	15.42	213.5
7.0	18.58	201.1	18.30	213.5	17.62	211.9
5.0	20.58	208.2	20.40	214.4	19.68	209.7
3.0	23.74	216.6	23.63	216.8	22.75	204.8
1.0	30.94	232.8	30.69	224.6	29.23	202.1
0.7	33.41	241.2	33.07	229.6	31.36	207.2
0.5	35.83	248.3	35.36	235.0	33.44	213.1
0.3	39.61	257.1	38.94	244.7	36.70	223.2
0.1	48.07	266.0	47.15	260.6	44.27	247.5
0.7E-1	50.84	265.0	49.86	257.8	46.89	254.5
0.5E-1	53.44	262.1	52.37	252.2	49.42	258.4
0.3E-1	57.31	254.8	56.07	241.8	53.29	257.7
0.1E-1	65.09	226.7	63.60	228.3	61.40	245.2
0.7E-2	67.41	218.2	65.96	225.2	63.94	240.6
0.5E-2	69.53	212.4	68.17	222.6	66.29	237.1
0.3E-2	72.67	208.4	71.48	220.6	69.80	232.8
0.1E-2	79.38	207.5	78.54	216.2	77.20	227.0
0.7E-3	81.53	205.4	80.76	214.8	79.54	226.2
0.5E-3	83.54	201.0	82.88	212.4	81.77	225.6
0.3E-3	86.48	193.0	86.01	205.2	85.14	222.5
0.1E-3	92.55	186.9	92.32	190.1	92.02	204.6
0.7E-4	94.53	187.0	94.32	189.4	94.15	201.5
0.5E-4	96.40	187.0	96.21	189.3	96.15	200.7
0.3E-4	99.25	187.6	99.10	190.8	99.21	203.1
0.1E-4	105.70	205.1	105.67	209.5	106.26	225.5

March

Geopotential height (km)	Pressure (kPa) 10 deg N	Temp (K)	Pressure (kPa) 40 deg N	Temp (K)	Pressure (kPa) 70 deg N	Temp (K)
0	101.10	300.4	101.58	284.9	101.58	256.4
1	90.14	294.7	89.92	277.8	88.90	255.5
2	80.19	289.2	79.39	271.4	77.67	252.9
3	71.19	283.7	69.91	265.7	67.75	248.8
4	63.07	278.4	61.40	260.6	58.98	243.4
5	55.73	272.8	53.76	254.2	51.16	237.3
6	49.13	266.9	46.93	247.2	44.22	230.7
7	43.19	260.6	40.80	240.0	38.07	224.5
8	37.85	254.1	35.33	232.8	32.65	219.9
9	33.03	247.0	30.43	226.5	27.92	217.3
10	28.70	239.3	26.13	221.7	23.86	217.2
11	24.83	231.4	22.37	218.8	20.39	217.7
12	21.37	223.1	19.13	216.9	17.44	217.7
13	18.28	214.6	16.34	216.0	14.91	217.7
14	15.54	206.6	13.94	215.5	12.75	217.7
15	13.12	199.9	11.90	215.2	10.91	217.6
16	11.03	196.3	10.15	214.9	9.326	217.4
17	9.253	196.2	8.658	214.8	7.973	217.2
18	7.771	198.9	7.385	214.8	6.816	217.0
19	6.546	202.4	6.299	214.9	5.826	217.0
20	5.528	205.6	5.373	215.2	4.979	217.1
21	4.681	208.6	4.585	215.6	4.258	217.5
22	3.975	211.5	3.915	216.2	3.641	218.0
23	3.383	214.3	3.344	216.9	3.115	218.4
24	2.885	216.9	2.857	217.7	2.664	218.6
25	2.466	219.2	2.443	218.5	2.279	218.8
30	1.153	230.3	1.131	225.7	1.050	222.6
35	0.5637	247.3	0.5437	241.2	0.4951	232.0
40	0.2886	262.5	0.2739	256.4	0.2413	243.5
45	0.1526	271.6	0.1428	266.2	0.1221	257.1
50	0.8141E-01	270.3	0.7519E-01	264.1	0.6335E-01	261.0
55	0.4278E-01	259.0	0.3879E-01	250.6	0.3267E-01	252.8
60	0.2162E-01	241.5	0.1921E-01	236.7	0.1630E-01	238.9
65	0.1043E-01	227.0	0.9226E-02	229.1	0.7849E-02	229.3
70	0.4812E-02	215.1	0.4336E-02	223.3	0.3695E-02	224.1
75	0.2147E-02	209.9	0.1996E-02	217.1	0.1709E-02	219.0
80	0.9527E-03	208.4	0.8982E-03	208.6	0.7768E-03	210.8
85	0.4137E-03	199.1	0.3872E-03	196.6	0.3321E-03	190.6
90	0.1700E-03	188.3	0.1568E-03	183.6	0.1291E-03	175.4
95	0.6878E-04	186.7	0.6158E-04	181.1	0.4907E-04	178.7
100	0.2804E-04	188.1	0.2472E-04	189.1	0.2004E-04	199.0
105	0.1189E-04	199.8	0.1081E-04	211.5	0.9341E-05	234.2

June

Geopotential height (km)	Pressure (kPa) 10 deg N	Temp (K)	Pressure (kPa) 40 deg N	Temp (K)	Pressure (kPa) 70 deg N	Temp (K)
0	101.05	301.1	101.69	290.3	101.16	276.4
1	90.12	295.2	90.35	287.1	89.34	273.6
2	80.19	289.5	80.11	283.1	78.74	269.7
3	71.20	283.8	70.89	278.1	69.26	264.9
4	63.07	278.3	62.62	272.4	60.80	259.0
5	55.74	272.6	55.15	266.1	53.19	252.8
6	49.14	266.7	48.44	259.4	46.40	246.1
7	43.19	260.6	42.39	252.3	40.31	239.0
8	37.86	254.3	36.97	245.0	34.88	232.4
9	33.04	247.2	32.08	237.9	30.06	228.0
10	28.72	239.5	27.71	231.0	25.87	226.3
11	24.84	231.5	23.85	224.9	22.25	227.3
12	21.39	223.1	20.46	220.1	19.14	227.4
13	18.30	214.8	17.49	217.0	16.47	227.6
14	15.56	207.3	14.92	214.9	14.17	227.8
15	13.15	201.0	12.71	213.7	12.19	227.9
16	11.06	197.9	10.82	213.1	10.49	228.1
17	9.302	198.3	9.216	213.4	9.031	228.5
18	7.828	201.6	7.849	214.3	7.775	229.0
19	6.611	205.4	6.691	215.5	6.696	229.3
20	5.597	208.8	5.708	216.7	5.768	229.6
21	4.751	211.8	4.873	218.0	4.969	229.9
22	4.044	214.3	4.166	219.1	4.281	229.7
23	3.448	216.7	3.564	220.4	3.688	229.4
24	2.946	219.1	3.052	221.9	3.177	229.8
25	2.522	221.3	2.618	223.6	2.739	230.4
30	1.185	230.9	1.241	234.0	1.321	239.8
35	0.5774	244.3	0.6098	247.4	0.6623	255.4
40	0.2917	255.3	0.3118	261.5	0.3465	271.0
45	0.1512	263.5	0.1647	271.8	0.1870	281.2
50	0.7932E-01	264.4	0.8778E-01	269.2	0.1022	282.0
55	0.4117E-01	255.3	0.4600E-01	258.5	0.5534E-01	273.9
60	0.2071E-01	240.8	0.2330E-01	243.3	0.2921E-01	259.4
65	0.9890E-02	220.8	0.1127E-01	226.8	0.1475E-01	240.5
70	0.4429E-02	206.2	0.5159E-02	210.3	0.7026E-02	220.1
75	0.1915E-02	202.7	0.2222E-02	196.2	0.3089E-02	195.1
80	0.8270E-03	202.0	0.9048E-03	183.3	0.1204E-02	167.0
85	0.3508E-03	194.3	0.3469E-03	174.0	0.3921E-03	142.6
90	0.1429E-03	187.1	0.1277E-03	169.0	0.1163E-03	144.6
95	0.5765E-04	186.5	0.4769E-04	177.2	0.3909E-04	170.2
100	0.2362E-04	190.4	0.1947E-04	198.7	0.1629E-04	216.4
105	0.1028E-04	206.6	0.9085E-05	234.2	0.8321E-05	270.0

September

Geopotential height (km)	Pressure (kPa) 10 deg N	Temp (K)	Pressure (kPa) 40 deg N	Temp (K)	Pressure (kPa) 70 deg N	Temp (K)
0	101.00	301.0	101.71	294.0	101.17	275.8
1	90.07	294.9	90.43	288.6	89.26	271.3
2	80.13	289.1	80.23	283.2	78.58	266.5
3	71.14	283.4	71.03	277.9	69.02	261.5
4	63.01	278.1	62.74	272.5	60.48	256.1
5	55.68	272.5	55.27	266.5	52.83	250.0
6	49.08	266.7	48.56	260.0	46.01	243.3
7	43.14	260.6	42.52	253.1	39.90	236.1
8	37.82	254.1	37.10	245.8	34.46	229.5
9	33.00	247.0	32.21	238.7	29.63	224.7
10	28.68	239.1	27.85	231.9	25.44	222.9
11	24.80	231.0	23.99	225.8	21.82	223.4
12	21.35	222.8	20.58	220.6	18.72	223.5
13	18.26	214.7	17.59	216.7	16.06	223.7
14	15.53	207.5	15.00	213.8	13.78	224.0
15	13.13	201.7	12.77	212.0	11.83	224.1
16	11.05	198.8	10.85	211.1	10.15	224.1
17	9.300	199.2	9.223	211.6	8.713	223.9
18	7.833	202.3	7.843	213.0	7.477	223.6
19	6.618	205.9	6.678	214.5	6.415	223.3
20	5.606	209.1	5.692	216.0	5.502	223.0
21	4.759	211.4	4.856	217.3	4.717	222.6
22	4.047	213.1	4.147	218.2	4.044	222.1
23	3.447	214.7	3.544	219.0	3.466	221.8
24	2.940	216.6	3.031	220.1	2.970	221.8
25	2.512	218.6	2.595	221.1	2.546	221.8
30	1.173	230.2	1.214	229.0	1.187	226.8
35	0.5704	243.8	0.5857	240.0	0.5667	235.0
40	0.2890	258.1	0.2933	253.8	0.2786	246.4
45	0.1512	267.6	0.1518	263.6	0.1418	258.6
50	0.7997E-01	267.0	0.7970E-01	264.0	0.7379E-01	261.9
55	0.4171E-01	256.2	0.4111E-01	250.7	0.3806E-01	252.9
60	0.2096E-01	240.2	0.2041E-01	237.1	0.1911E-01	242.8
65	0.1008E-01	226.3	0.9758E-02	225.5	0.9314E-02	231.7
70	0.4638E-02	214.4	0.4496E-02	215.8	0.4366E-02	219.2
75	0.2061E-02	208.0	0.2012E-02	209.8	0.1961E-02	208.5
80	0.9064E-03	206.2	0.8854E-03	203.8	0.8510E-03	198.0
85	0.3905E-03	197.4	0.3732E-03	190.4	0.3395E-03	174.6
90	0.1598E-03	187.8	0.1467E-03	178.4	0.1214E-03	163.9
95	0.6460E-04	186.6	0.5652E-04	179.1	0.4424E-04	175.0
100	0.2637E-04	188.8	0.2270E-04	191.4	0.1814E-04	203.8
105	0.1130E-04	202.9	0.1014E-04	219.2	0.8721E-05	246.5

December

Geopotential height (km)	Pressure (kPa) 10 deg N	Temp (K)	Pressure (kPa) 40 deg N	Temp (K)	Pressure (kPa) 70 deg N	Temp (K)
0	101.05	300.6	101.59	286.1	101.22	257.2
1	90.08	294.4	89.96	278.0	88.60	255.6
2	80.13	288.5	79.44	271.1	77.43	252.8
3	71.13	283.1	69.96	265.2	67.56	248.9
4	63.00	278.1	61.43	260.4	58.83	243.7
5	55.67	272.6	53.80	254.2	51.05	237.7
6	49.07	266.7	46.97	247.4	44.14	231.1
7	43.13	260.4	40.85	240.3	38.02	224.7
8	37.80	253.7	35.38	233.2	32.61	219.8
9	32.98	246.5	30.49	227.0	27.88	216.6
10	28.65	238.7	26.19	222.0	23.81	215.7
11	24.77	230.6	22.42	218.7	20.32	215.5
12	21.31	222.3	19.17	216.3	17.32	215.2
13	18.22	213.7	16.36	214.9	14.76	214.7
14	15.48	205.7	13.95	214.0	12.57	214.2
15	13.06	199.0	11.89	213.5	10.70	213.7
16	10.97	195.7	10.13	213.2	9.102	213.1
17	9.207	196.0	8.624	213.2	7.740	212.4
18	7.734	199.1	7.346	213.4	6.579	211.6
19	6.518	202.8	6.259	213.8	5.588	210.6
20	5.508	206.2	5.334	214.2	4.742	209.3
21	4.667	209.4	4.548	214.8	4.018	207.6
22	3.966	212.1	3.880	215.7	3.402	206.0
23	3.377	214.8	3.312	216.4	2.877	204.4
24	2.881	217.1	2.828	217.0	2.432	202.5
25	2.463	219.0	2.417	217.5	2.053	201.0
30	1.149	229.8	1.111	223.2	0.8782	203.8
35	0.5606	246.0	0.5267	234.2	0.3903	217.9
40	0.2849	257.9	0.2591	247.8	0.1832	234.0
45	0.1484	264.5	0.1325	259.7	0.9042E-01	249.6
50	0.7801E-01	265.6	0.6868E-01	257.5	0.4632E-01	258.7
55	0.4077E-01	259.6	0.3485E-01	244.9	0.2390E-01	256.0
60	0.2081E-01	246.8	0.1703E-01	233.1	0.1214E-01	247.8
65	0.1013E-01	227.1	0.8097E-02	226.5	0.6021E-02	238.9
70	0.4637E-02	211.4	0.3772E-02	221.3	0.2915E-02	232.5
75	0.2045E-02	207.6	0.1735E-02	218.6	0.1389E-02	227.9
80	0.9026E-03	207.1	0.7904E-03	215.4	0.6529E-03	226.1
85	0.3896E-03	196.8	0.3550E-03	208.1	0.3063E-03	222.7
90	0.1590E-03	187.6	0.1511E-03	193.2	0.1397E-03	209.3
95	0.6432E-04	187.0	0.6203E-04	189.3	0.6071E-04	200.9
100	0.2625E-04	188.0	0.2560E-04	191.7	0.2633E-04	204.4
105	0.1118E-04	201.1	0.1111E-04	206.0	0.1201E-04	219.8

6. Mean reference atmosphere (110–500 km) (for average solar flux and magnetically quiet conditions; exospheric temperature 1037 K)

Geometric height (km)	Temp (K)	Pressure (Nm^{-2})	Mean mol. mass
110	232	6.3×10^{-3}	27.3
120	377	2.2×10^{-3}	26.2
130	518	1.1×10^{-3}	25.3
140	628	6.9×10^{-4}	24.6
150	715	4.6×10^{-4}	24.0
160	783	3.2×10^{-4}	23.5
170	837	2.3×10^{-4}	23.0
180	879	1.7×10^{-4}	22.5
190	912	1.3×10^{-4}	22.0
200	938	1.0×10^{-4}	21.6
250	1005	3.3×10^{-5}	19.5
300	1027	1.2×10^{-5}	18.1
350	1034	4.7×10^{-6}	17.1
400	1036	2.0×10^{-6}	16.4
450	1037	8.8×10^{-7}	15.8
500	1037	4.2×10^{-7}	15.1

From COSPAR, 1988.

7. The Planck function

$$B_{\tilde{\nu}} = \frac{c_1 \tilde{\nu}^3}{\exp(c_2 \tilde{\nu}/T) - 1}$$

where $B_{\tilde{\nu}}$ is radiance (in $Wm^{-2}sr^{-1}(cm^{-1})^{-1}$) of black body at T K and $\tilde{\nu}$ wavenumbers (cm^{-1}), c_1 and c_2 are known as first and second radiation constants and have the values

$$c_1 = 1.1911 \times 10^{-8} \, Wm^{-2}sr^{-1}(cm^{-1})^{-4}$$

$$c_2 = 1.439 K(cm^{-1})^{-1}$$

$$B_\lambda = \frac{c_1}{\lambda^5(\exp(c_2/\lambda T) - 1)}$$

where B_λ is radiance in $Wm^{-2}sr^{-1}cm^{-1}$ of black body at T K and wavelength λ cm, c_1 and c_2 have the same values as above.

Another useful quantity is

$$D = \frac{\int_{\tilde{\nu}}^{\infty} B_{\tilde{\nu}} d\tilde{\nu}}{\int_0^{\infty} B_{\tilde{\nu}} d\tilde{\nu}}$$

A series evaluation of D suitable for numerical computation is

$$D = \frac{15}{\pi^4} \sum_{m=1}^{\infty} m^{-4} \exp(-mv)[\{(mv+3)mv+6\}mv+6]$$

For small values of v, $v < 2\pi$, the following series converges more rapidly

$$D = 1 - \frac{15}{\pi^4} v^3 \left(\frac{1}{3} - \frac{v}{8} + \frac{v^2}{60} - \frac{v^4}{5040} + \frac{v^6}{272\,160} - \frac{v^8}{13\,305\,600} \right)$$

where

$$v = \frac{c_2 \tilde{\nu}}{T}$$

The wavenumber $\tilde{\nu}_m\,\mathrm{cm}^{-1}$ of maximum $B_{\tilde{\nu}}$ is given by Wien's displacement law which is

$$\tilde{\nu}_m = 1.9609\,T$$

The wavelength $\lambda_m\,\mathrm{cm}$ of maximum B_λ is given by

$$\lambda_m = 0.289\,79/T.$$

Reference Pivovonsky & Nagel (1961).

8.   Solar radiation

Solar constant (i.e. mean value of total solar radiation incident on surface just outside earth's atmosphere normal to solar beam) = 1370 W m^{-2}.

Fig. A8.1. Solar irradiation curves. Shaded areas show absorption by vertical path of whole atmosphere by constituents shown. (From Air Force Cambridge Research Laboratories, 1965)

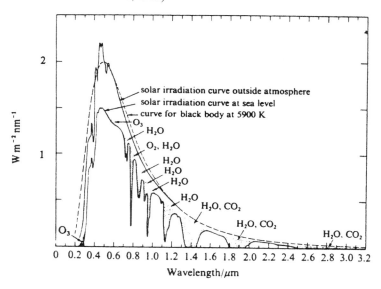

Spectral details of the solar irradiance have been published by Kurucz (1995). Information can also be found on the HITRAN Molecular Spectroscopic Database, 1996 edn (*http://www.hitran.com*) available as a CD-ROM.

9.   Absorption by ozone in the ultraviolet

Table A9. *Absorption cross-sections by* O_3 *in* cm^2/molecule.

Wavelength (nm)	cm^2/molecule
200	3.1540E-19
210	5.7160E-19
220	1.7850E-18
230	4.4760E-18
240	8.3119E-18
250	1.1146E-17
260	1.0708E-17
270	7.8556E-18
280	3.8834E-18
290	1.3369E-18
300	3.6380E-19
310	8.7162E-20
320	2.8538E-20
330	3.2276E-21
340	1.6102E-21
450	1.7010E-22
500	1.1760E-21
550	3.3140E-21
600	5.0694E-21
650	2.5000E-21
700	8.7454E-22
750	4.2745E-22

Data from Anderson & Mauersberger (1992), Brion et al. (1993), and Malicet et al. (1995).

10.   Spectral band information

The following tables list values of Σs_i (columns S) and $\Sigma(s_i \gamma_{oi})^{1/2}$ (columns R) at different temperatures over different wavenumber intervals of various infrared bands of water vapour, carbon dioxide and ozone, where s_i is the strength of the ith line in $cm^{-1} (g\,cm^{-2})^{-1}$ and γ_{oi} is the collision broadened half-width in cm^{-1} of the ith line at standard pressure (101.3 kPa). The data are from the HITRAN Molecular Spectroscopic Database: 1996 edn (http://www.hitran.com). More explanation is given in chapter 4.

In the tables, each number is given by a number between 1 and 10 expressed to four significant figures followed by an exponent of ten, e.g. the value for S at 220 K for water vapour between 0 and 25 cm^{-1} is 3.627×10^3.

Water vapour

Temperature		220 K		260 K		300 K	
cm-1	interval	S	R	S	R	S	R
0	25	3.627e + 03	3.342e + 01	2.496e + 03	2.778e + 01	1.806e + 03	2.377e + 01
25	50	3.714e + 04	2.199e + 02	2.980e + 04	1.902e + 02	2.431e + 04	1.673e + 02
50	75	1.190e + 05	4.470e + 02	9.409e + 04	3.951e + 02	7.644e + 04	3.538e + 02
75	100	2.179e + 05	5.639e + 02	1.772e + 05	5.134e + 02	1.472e + 05	4.728e + 02
100	125	1.977e + 05	5.046e + 02	1.859e + 05	4.937e + 02	1.715e + 05	4.794e + 02
125	150	2.295e + 05	5.109e + 02	2.070e + 05	4.899e + 02	1.868e + 05	4.699e + 02
150	175	2.958e + 05	5.013e + 02	2.761e + 05	4.961e + 02	2.558e + 05	4.885e + 02
175	200	7.963e + 04	2.510e + 02	9.467e + 04	2.780e + 02	1.081e + 05	2.996e + 02
200	225	2.452e + 05	4.433e + 02	2.465e + 05	4.420e + 02	2.399e + 05	4.357e + 02
225	250	1.260e + 05	2.779e + 02	1.411e + 05	2.959e + 02	1.497e + 05	3.086e + 02
250	275	7.903e + 04	1.637e + 02	9.529e + 04	1.792e + 02	1.048e + 05	1.896e + 02
275	300	6.294e + 04	2.233e + 02	8.510e + 04	2.514e + 02	1.025e + 05	2.704e + 02
300	325	4.339e + 04	1.612e + 02	6.575e + 04	1.911e + 02	8.633e + 04	2.148e + 02
325	350	1.868e + 04	1.129e + 02	3.194e + 04	1.382e + 02	4.646e + 04	1.590e + 02
350	375	9.610e + 03	8.717e + 01	1.819e + 04	1.152e + 02	2.929e + 04	1.404e + 02
375	400	4.184e + 03	6.188e + 01	8.588e + 03	8.422e + 01	1.530e + 04	1.065e + 02
400	425	2.891e + 03	3.523e + 01	4.963e + 03	4.935e + 01	7.879e + 03	6.417e + 01
425	450	7.203e + 02	2.400e + 01	1.320e + 03	3.289e + 01	2.682e + 03	4.407e + 01
450	475	1.445e + 03	3.494e + 01	2.551e + 03	4.778e + 01	4.015e + 03	6.118e + 01
475	500	3.173e + 02	9.724e + 00	4.779e + 02	1.417e + 01	7.880e + 02	2.019e + 01
500	525	5.260e + 02	1.657e + 01	1.050e + 03	2.414e + 01	1.808e + 03	3.287e + 01
525	550	2.968e + 02	1.099e + 01	5.247e + 02	1.451e + 01	7.971e + 02	1.856e + 01
550	575	1.151e + 02	7.615e + 00	1.813e + 02	9.960e + 00	2.632e + 02	1.289e + 01
575	600	1.560e + 02	8.577e + 00	3.526e + 02	1.269e + 01	6.346e + 02	1.693e + 01
600	625	4.551e + 01	4.078e + 00	9.785e + 01	6.021e + 00	1.869e + 02	8.356e + 00
625	650	5.306e + 01	4.967e + 00	1.295e + 02	7.398e + 00	2.527e + 02	1.002e + 01
650	675	2.442e + 01	3.290e + 00	4.569e + 01	4.611e + 00	7.774e + 01	6.098e + 00

Water vapour (continued)

Temperature		220 K		260 K		300 K	
cm-1	interval	S	R	S	R	S	R
675	700	1.612e+01	2.812e+00	4.696e+01	4.556e+00	1.028e+02	6.469e+00
700	725	9.256e+00	2.001e+00	2.271e+01	3.212e+00	4.771e+01	4.642e+00
725	750	5.356e+00	1.558e+00	1.422e+01	2.500e+00	3.184e+01	3.621e+00
750	775	1.542e+00	7.257e-01	4.117e+00	1.214e+00	9.935e+00	1.870e+00
775	800	5.139e+00	1.588e+00	1.303e+01	2.548e+00	2.680e+01	3.626e+00
1200	1250	2.006e+01	6.088e+00	3.961e+01	8.887e+00	6.989e+01	1.236e+01
1250	1300	9.144e+01	1.680e+01	1.879e+02	2.423e+01	3.537e+02	3.312e+01
1300	1350	5.893e+02	4.530e+01	1.233e+03	6.071e+01	2.176e+03	7.676e+01
1350	1400	5.022e+03	9.987e+01	7.830e+03	1.217e+02	1.077e+04	1.406e+02
1400	1450	1.044e+04	1.348e+02	1.365e+04	1.523e+02	1.644e+04	1.667e+02
1450	1500	3.154e+04	2.673e+02	3.494e+04	2.804e+02	3.715e+04	2.911e+02
1500	1550	6.472e+04	4.957e+02	6.587e+04	5.121e+02	6.653e+04	5.258e+02
1550	1600	4.685e+04	2.736e+02	4.285e+04	2.542e+02	3.925e+04	2.383e+02
1600	1650	4.023e+04	2.692e+02	3.469e+04	2.444e+02	3.031e+04	2.261e+02
1650	1700	7.894e+04	4.427e+02	7.153e+04	4.167e+02	6.523e+04	3.953e+02
1700	1750	3.926e+04	3.165e+02	4.071e+04	3.208e+02	4.119e+04	3.224e+02
1750	1800	2.093e+04	1.874e+02	2.321e+04	2.003e+02	2.525e+04	2.108e+02
1800	1850	8.125e+03	1.038e+02	9.438e+03	1.155e+02	1.070e+04	1.262e+02
1850	1900	2.952e+03	5.231e+01	3.562e+03	5.875e+01	4.056e+03	6.505e+01
1900	1950	1.984e+03	4.510e+01	2.703e+03	5.231e+01	3.325e+03	5.857e+01
1950	2000	4.363e+02	2.137e+01	7.288e+02	2.659e+01	1.041e+03	3.141e+01
2000	2050	1.216e+02	1.181e+01	2.383e+02	1.638e+01	3.875e+02	2.099e+01
2050	2100	4.368e+01	6.818e+00	9.837e+01	1.036e+01	1.785e+02	1.426e+01
2800	2900	5.470e+00	1.001e+01	6.548e+00	1.062e+01	7.852e+00	1.121e+01
2900	3000	4.397e+01	1.403e+01	7.134e+01	1.797e+01	1.028e+02	2.181e+01
3000	3100	5.190e+02	4.811e+01	5.527e+02	4.987e+01	5.735e+02	5.125e+01
3100	3200	5.937e+02	5.119e+01	5.503e+02	5.169e+01	5.203e+02	5.304e+01

3200	3300	7.208e+02	6.214e+01	6.987e+02	6.414e+01	6.948e+02	6.690e+01
3300	3400	4.180e+02	5.052e+01	5.709e+02	6.113e+01	7.544e+02	7.143e+01
3400	3500	9.971e+02	7.845e+01	1.418e+03	9.790e+01	1.910e+03	1.179e+02
3500	3600	8.171e+03	2.852e+02	1.174e+04	3.326e+02	1.571e+04	3.733e+02
3600	3700	6.216e+04	5.730e+02	6.455e+04	5.707e+02	6.516e+04	5.642e+02
3700	3800	7.820e+04	6.861e+02	7.363e+04	6.718e+02	7.008e+04	6.603e+02
3800	3900	1.016e+05	7.022e+02	9.605e+04	6.775e+02	9.048e+04	6.531e+02
3900	4000	7.171e+03	1.690e+02	1.054e+04	2.041e+02	1.405e+04	2.348e+02
4000	4100	3.622e+02	3.647e+01	4.312e+02	4.123e+01	5.042e+02	4.645e+01
4100	4200	7.379e+01	1.875e+01	1.097e+02	2.154e+01	1.468e+02	2.392e+01
4200	4300	6.415e+00	7.035e+00	1.260e+01	8.741e+00	2.167e+01	1.045e+01
4300	4400	4.034e-01	1.209e+00	1.036e+00	1.675e+00	2.211e+00	2.208e+00
4400	4500	8.107e-02	3.565e-01	1.676e-01	4.726e-01	2.882e-01	5.818e-01
4500	4600	2.624e+00	2.931e+00	3.017e+00	3.088e+00	3.326e+00	3.176e+00
4600	4700	1.934e+00	2.384e+00	1.789e+00	2.366e+00	1.689e+00	2.362e+00
4700	4800	2.026e+00	2.730e+00	1.966e+00	2.778e+00	1.950e+00	2.864e+00
4800	4900	1.575e+00	3.012e+00	1.969e+00	3.639e+00	2.615e+00	4.297e+00
4900	5000	4.824e+00	6.148e+00	8.409e+00	7.836e+00	1.299e+01	9.429e+00
5000	5100	4.329e+01	1.685e+01	6.402e+01	2.036e+01	9.411e+01	2.410e+01
5100	5200	8.292e+02	7.732e+01	1.226e+03	9.011e+01	1.642e+03	1.014e+02
5200	5300	7.139e+03	1.801e+02	7.078e+03	1.759e+02	6.898e+03	1.710e+02
5300	5400	1.091e+04	2.302e+02	9.944e+03	2.225e+02	9.222e+03	2.168e+02
5400	5500	8.970e+03	2.046e+02	9.163e+03	2.028e+02	9.137e+03	1.995e+02
5500	5600	4.770e+02	4.641e+01	7.748e+02	5.717e+01	1.119e+03	6.654e+01
5600	5700	4.310e+01	1.197e+01	5.365e+01	1.456e+01	6.806e+01	1.746e+01
5700	5800	8.569e+00	4.481e+00	1.188e+01	5.371e+00	1.507e+01	6.184e+00
5800	5900	1.562e+00	1.785e+00	2.822e+00	2.353e+00	4.358e+00	2.876e+00
5900	6000	1.728e-01	5.914e-01	4.090e-01	8.893e-01	7.968e-01	1.208e+00
6000	6100	2.122e-01	9.325e-01	2.687e-01	1.058e+00	3.533e-01	1.192e+00
6100	6200	5.565e-02	3.801e-01	5.930e-02	3.956e-01	6.756e-02	4.181e-01
6200	6300	9.679e-02	5.887e-01	1.032e-01	5.946e-01	1.083e-01	5.962e-01
6300	6400	7.334e-02	5.533e-01	9.593e-02	6.366e-01	1.247e-01	7.130e-01

Water vapour (*continued*)

Temperature		220 K		260 K		300 K	
cm-1	interval	S	R	S	R	S	R
6400	6500	1.039e − 01	6.768e − 01	1.949e − 01	9.134e − 01	3.398e − 01	1.161e + 00
6500	6600	1.464e + 00	2.696e + 00	2.481e + 00	3.582e + 00	3.730e + 00	4.451e + 00
6600	6700	1.525e + 01	1.010e + 01	2.009e + 01	1.203e + 01	2.614e + 01	1.400e + 01
6700	6800	2.154e + 02	3.608e + 01	2.435e + 02	3.892e + 01	2.654e + 02	4.125e + 01
6800	6900	3.786e + 02	4.711e + 01	3.451e + 02	4.665e + 01	3.226e + 02	4.702e + 01
6900	7000	6.046e + 02	6.873e + 01	6.256e + 02	7.392e + 01	6.627e + 02	7.920e + 01
7000	7100	9.764e + 02	9.979e + 01	1.381e + 03	1.157e + 02	1.771e + 03	1.281e + 02
7100	7200	4.916e + 03	1.732e + 02	4.999e + 03	1.739e + 02	4.980e + 03	1.738e + 02
7200	7300	7.007e + 03	2.184e + 02	6.436e + 03	2.087e + 02	5.983e + 03	2.004e + 02
7300	7400	7.853e + 03	2.148e + 02	7.906e + 03	2.155e + 02	7.843e + 03	2.140e + 02
7400	7500	2.027e + 02	4.055e + 01	3.092e + 02	4.770e + 01	4.459e + 02	5.473e + 01
7500	7600	4.092e + 01	1.729e + 01	4.491e + 01	1.821e + 01	4.837e + 01	1.893e + 01
7600	7700	1.147e + 01	7.595e + 00	1.471e + 01	8.653e + 00	1.783e + 01	9.478e + 00
7700	7800	8.715e − 01	2.127e + 00	1.452e + 00	2.802e + 00	2.264e + 00	3.470e + 00
7800	7900	1.424e − 01	6.419e − 01	2.323e − 01	8.147e − 01	3.444e − 01	9.765e − 01
7900	8000	3.304e − 02	3.022e − 01	6.263e − 02	4.001e − 01	1.029e − 01	4.899e − 01
8000	8100	7.844e − 02	3.818e − 01	1.160e − 01	4.472e − 01	1.672e − 01	5.080e − 01
8100	8200	1.757e + 00	2.834e + 00	2.037e + 00	3.009e + 00	2.284e + 00	3.136e + 00
8200	8300	5.674e + 00	5.767e + 00	6.298e + 00	5.910e + 00	6.799e + 00	5.979e + 00
8300	8400	1.071e + 01	8.297e + 00	9.954e + 00	7.888e + 00	9.375e + 00	7.568e + 00
8400	8500	1.762e + 01	1.020e + 01	1.668e + 01	9.859e + 00	1.591e + 01	9.575e + 00
8500	8600	1.049e + 01	1.052e + 01	1.473e + 01	1.219e + 01	2.041e + 01	1.376e + 01
8600	8700	1.124e + 02	2.757e + 01	1.449e + 02	3.030e + 01	1.727e + 02	3.212e + 01
8700	8800	2.728e + 02	3.691e + 01	2.535e + 02	3.466e + 01	2.349e + 02	3.267e + 01
8800	8900	6.020e + 02	6.547e + 01	5.584e + 02	6.222e + 01	5.220e + 02	5.940e + 01
8900	9000	2.525e + 02	4.129e + 01	2.889e + 02	4.329e + 01	3.164e + 02	4.438e + 01
9000	9100	1.442e + 01	1.273e + 01	1.703e + 01	1.332e + 01	2.123e + 01	1.403e + 01

9100	9200	3.974e + 00	4.668e + 00	4.284e + 00	4.784e + 00	4.541e + 00	4.827e + 00
9200	9300	1.485e + 00	2.033e + 00	1.768e + 00	2.199e + 00	1.996e + 00	2.302e + 00
9300	9400	1.660e − 01	6.145e − 01	2.792e − 01	7.660e − 01	4.141e − 01	8.948e − 01
9400	9500	8.181e − 03	7.610e − 02	1.424e − 02	1.024e − 01	2.224e − 02	1.276e − 01
9500	9600	0.000e + 00	0.000e + 00	0.000e + 00	0.000e + 00	0.000e + 00	0.000e + 00
9600	9700	1.156e − 02	1.028e − 01	1.766e − 02	1.233e − 01	2.405e − 02	1.393e − 01
9700	9800	1.562e − 01	6.061e − 01	1.635e − 01	6.014e − 01	1.673e − 01	5.891e − 01
9800	9900	2.072e − 01	7.238e − 01	1.897e − 01	6.743e − 01	1.769e − 01	6.333e − 01
9900	10000	3.897e − 01	1.236e + 00	3.930e − 01	1.233e + 00	3.943e − 01	1.220e + 00
10000	10100	2.266e − 01	9.844e − 01	3.249e − 01	1.165e + 00	4.495e − 01	1.327e + 00
10100	10200	2.476e + 00	3.895e + 00	3.445e + 00	4.519e + 00	4.430e + 00	5.029e + 00
10200	10300	1.628e + 01	1.011e + 01	1.664e + 01	1.039e + 01	1.695e + 01	1.063e + 01
10300	10400	2.997e + 01	1.616e + 01	3.150e + 01	1.698e + 01	3.447e + 01	1.772e + 01
10400	10500	8.706e + 01	2.945e + 01	1.037e + 02	3.113e + 01	1.167e + 02	3.206e + 01
10500	10600	2.036e + 02	3.879e + 01	2.029e + 02	3.832e + 01	2.006e + 02	3.763e + 01
10600	10700	2.682e + 02	4.393e + 01	2.412e + 02	4.134e + 01	2.206e + 02	3.918e + 01
10700	10800	9.826e + 01	2.736e + 01	1.153e + 02	2.966e + 01	1.295e + 02	3.132e + 01
10800	10900	1.042e + 01	9.719e + 00	1.136e + 01	9.846e + 00	1.220e + 01	9.844e + 00
10900	11000	2.248e + 01	1.465e + 01	2.238e + 01	1.443e + 01	2.219e + 01	1.411e + 01
11000	11100	2.692e + 01	1.142e + 01	2.403e + 01	1.057e + 01	2.165e + 01	9.836e + 00
11100	11200	1.866e + 01	9.276e + 00	1.987e + 01	9.544e + 00	2.072e + 01	9.658e + 00

Carbon dioxide

425	450	1.259e − 05	2.472e − 03	1.252e − 04	7.370e − 03	6.461e − 04	1.593e − 02
450	475	3.505e − 04	3.314e − 02	2.624e − 03	8.617e − 02	1.111e − 02	1.692e − 01
475	500	5.097e − 04	5.325e − 02	4.183e − 03	1.508e − 01	1.937e − 02	3.183e − 01
500	525	2.105e − 02	3.027e − 01	1.089e − 01	7.102e − 01	3.595e − 01	1.316e + 00
525	550	3.338e − 01	1.360e + 00	1.275e + 00	2.700e + 00	3.357e + 00	4.507e + 00
550	575	5.544e − 01	2.489e + 00	2.646e + 00	5.444e + 00	8.429e + 00	9.759e + 00
575	600	5.469e + 01	2.097e + 01	1.328e + 02	3.519e + 01	2.544e + 02	5.216e + 01
600	625	5.513e + 02	6.365e + 01	1.011e + 03	8.877e + 01	1.555e + 03	1.148e + 02

Carbon dioxide (*continued*)

Temperature		220K		260K		300K	
cm-1	interval	S	R	S	R	S	R
625	650	7.493e + 03	2.165e + 02	9.057e + 03	2.539e + 02	1.042e + 04	2.920e + 02
650	675	8.839e + 04	8.028e + 02	8.469e + 04	8.320e + 02	8.120e + 04	8.705e + 02
675	700	2.750e + 04	2.800e + 02	2.780e + 04	3.110e + 02	2.771e + 04	3.453e + 02
700	725	1.226e + 03	8.641e + 01	2.271e + 03	1.234e + 02	3.536e + 03	1.622e + 02
725	750	1.860e + 02	2.811e + 01	3.640e + 02	4.361e + 01	5.930e + 02	6.155e + 01
750	775	6.729e + 00	6.084e + 00	2.416e + 01	1.251e + 01	6.119e + 01	2.139e + 01
775	800	9.185e − 01	2.393e + 00	3.277e + 00	4.761e + 00	8.447e + 00	8.074e + 00
800	825	2.863e − 01	7.807e − 01	1.042e + 00	1.523e + 00	2.633e + 00	2.487e + 00
825	850	9.669e − 03	2.742e − 01	6.704e − 02	7.052e − 01	2.729e − 01	1.391e + 00
850	875	2.323e − 03	1.426e − 01	1.697e − 02	3.912e − 01	7.365e − 02	8.155e − 01
875	900	4.202e − 03	1.473e − 01	2.568e − 02	3.964e − 01	1.099e − 01	8.334e − 01
900	925	2.226e − 02	3.500e − 01	1.401e − 01	8.652e − 01	5.591e − 01	1.695e + 00
925	950	2.696e − 01	8.229e − 01	1.146e + 00	1.753e + 00	3.268e + 00	3.033e + 00
950	975	4.290e − 01	9.249e − 01	1.509e + 00	1.695e + 00	3.678e + 00	2.636e + 00
975	1000	1.619e − 01	5.251e − 01	7.358e − 01	1.098e + 00	2.194e + 00	1.865e + 00
1000	1025	1.111e − 02	1.722e − 01	6.248e − 02	3.893e − 01	2.405e − 01	7.340e − 01
1025	1050	3.434e − 01	8.240e − 01	1.374e + 00	1.711e + 00	3.752e + 00	2.939e + 00
1050	1075	7.844e − 01	1.490e + 00	2.533e + 00	2.760e + 00	5.865e + 00	4.357e + 00
1075	1100	4.612e − 01	1.188e + 00	1.866e + 00	2.519e + 00	5.198e + 00	4.385e + 00
1800	1825	2.330e − 04	1.955e − 02	1.212e − 03	4.283e − 02	3.986e − 02	7.460e − 02
1825	1850	2.668e − 03	7.072e − 02	1.003e − 02	1.328e − 01	2.616e − 01	2.076e − 01
1850	1875	4.144e − 02	4.387e − 01	8.301e − 02	6.243e − 01	1.429e − 01	8.219e − 01
1875	1900	2.754e − 01	1.569e + 00	5.333e − 01	2.018e + 00	9.030e − 01	2.477e + 00
1900	1925	4.776e + 00	3.925e + 00	5.206e + 00	4.195e + 00	5.505e + 00	4.409e + 00
1925	1950	3.078e + 00	3.188e + 00	2.968e + 00	3.143e + 00	2.863e + 00	3.129e + 00
1950	1975	2.704e − 01	7.136e − 01	3.741e − 01	8.852e − 01	4.793e − 01	1.052e + 00
1975	2000	2.387e − 02	3.403e − 01	5.648e − 02	4.886e − 01	1.073e − 01	6.363e − 01
2000	2025	3.569e − 01	1.043e + 00	4.218e − 01	1.239e + 00	5.036e − 01	1.452e + 00

2025	2050	4.324e + 00	4.537e + 00	6.650e + 00	5.134e + 00	9.219e + 00	5.720e + 00
2050	2075	3.482e + 01	9.760e + 00	3.555e + 01	1.032e + 01	3.591e + 01	1.095e + 01
2075	2100	3.559e + 01	1.383e + 01	3.588e + 01	1.500e + 01	3.660e + 01	1.627e + 01
2100	2125	1.079e + 00	2.772e + 00	2.129e + 00	4.132e + 00	3.548e + 00	5.630e + 00
2125	2150	6.028e − 01	1.726e + 00	1.182e + 00	2.631e + 00	1.971e + 00	3.669e + 00
2150	2175	2.417e − 02	4.986e − 01	8.629e − 02	9.677e − 01	2.263e − 01	1.588e + 00
2175	2200	1.523e − 02	3.813e − 01	8.852e − 02	9.795e − 01	3.476e − 01	2.013e + 00
2200	2225	9.818e − 01	3.182e + 00	4.400e + 00	6.870e + 00	1.367e + 01	1.232e + 01
2225	2250	2.283e + 02	3.202e + 01	4.107e + 02	4.611e + 01	6.374e + 02	6.212e + 01
2250	2275	4.684e + 03	1.209e + 02	4.819e + 03	1.391e + 02	4.909e + 03	1.671e + 02
2275	2300	8.127e + 03	2.228e + 02	9.217e + 03	2.874e + 02	1.226e + 04	3.761e + 02
2300	2325	1.069e + 05	6.413e + 02	1.462e + 05	7.932e + 02	1.840e + 05	9.417e + 02
2325	2350	5.633e + 05	1.303e + 03	5.317e + 05	1.327e + 03	5.016e + 05	1.366e + 03
2350	2375	6.813e + 05	1.239e + 03	6.617e + 05	1.241e + 03	6.395e + 05	1.249e + 03
2375	2400	1.244e + 04	9.341e + 01	2.242e + 04	1.271e + 02	3.403e + 04	1.590e + 02
2400	2425	4.848e − 02	2.700e − 01	1.688e − 01	4.817e − 01	4.105e − 01	7.250e − 01
2425	2450	5.346e − 02	3.737e − 01	1.789e − 01	6.524e − 01	4.239e − 01	9.740e − 01
2450	2475	1.248e − 02	2.920e − 01	3.949e − 02	4.613e − 01	1.072e − 01	6.711e − 01
2475	2500	9.320e − 02	6.195e − 01	9.096e − 02	6.232e − 01	9.013e − 02	6.374e − 01
3100	3200	8.647e − 02	7.005e − 01	9.751e − 02	7.717e − 01	1.095e − 01	8.326e − 01
3200	3300	6.187e − 02	5.883e − 01	9.074e − 02	7.195e − 01	1.260e − 01	8.423e − 01
3300	3400	1.194e + 00	3.278e + 00	1.375e + 00	3.730e + 00	1.561e + 00	4.179e + 00
3400	3500	4.980e + 00	9.044e + 00	8.214e + 00	1.265e + 01	1.235e + 01	1.713e + 01
3500	3600	4.312e + 03	2.001e + 02	4.905e + 03	2.359e + 02	5.463e + 03	2.741e + 02
3600	3700	1.586e + 04	3.493e + 02	1.585e + 04	3.646e + 02	1.576e + 04	3.848e + 02
3700	3800	1.752e + 04	2.965e + 02	1.694e + 04	3.171e + 02	1.653e + 04	3.416e + 02
3800	3900	1.012e − 01	9.352e − 01	2.980e − 01	1.411e + 00	7.015e − 01	2.018e + 00
3900	4000	1.041e − 02	2.986e − 01	1.711e − 02	3.729e − 01	2.917e − 02	4.559e − 01
4000	4100	1.730e − 02	2.811e − 01	3.446e − 02	3.701e − 01	5.884e − 02	4.569e − 01
4100	4200	0.000e + 00	0.000e + 00	0.000e + 00	0.000e + 00	0.000e + 00	0.000e + 00
4200	4300	0.000e + 00	0.000e + 00	0.000e + 00	0.000e + 00	0.000e + 00	0.000e + 00
4300	4400	0.000e + 00	0.000e + 00	0.000e + 00	0.000e + 00	0.000e + 00	0.000e + 00
4400	4500	1.284e − 03	4.168e − 02	1.411e − 03	4.133e − 02	1.488e − 03	4.027e − 02
4500	4600	1.102e − 02	2.625e − 01	1.391e − 02	2.903e − 01	1.709e − 02	3.113e − 01

Carbon dioxide (*continued*)

cm-1	interval	220K S	R	260K S	R	300K S	R
4600	4700	2.265e − 01	2.099e + 00	2.356e − 01	2.184e + 00	2.485e − 01	2.262e + 00
4700	4800	2.372e + 00	7.866e + 00	3.869e + 00	1.016e + 01	6.072e + 00	1.278e + 01
4800	4900	1.218e + 02	3.341e + 01	1.212e + 02	3.457e + 01	1.203e + 02	3.574e + 01
4900	5000	5.014e + 02	6.141e + 01	4.927e + 02	6.542e + 01	4.845e + 02	7.011e + 01
5000	5100	9.110e + 01	2.181e + 01	9.762e + 01	2.356e + 01	1.040e + 02	2.534e + 01
5100	5200	8.698e + 01	1.943e + 01	8.863e + 01	2.180e + 01	9.078e + 01	2.428e + 01
5200	5300	7.886e − 02	6.790e − 01	1.001e − 01	8.523e − 01	1.271e − 01	1.011e + 00
5300	5400	6.360e − 01	1.734e + 00	6.128e − 01	1.658e + 00	5.873e − 01	1.583e + 00

Ozone

cm-1	interval	220K S	R	260K S	R	300K S	R
0	25	5.929e + 02	1.743e + 02	4.770e + 02	1.597e + 02	3.921e + 02	1.485e + 02
25	50	2.070e + 03	3.715e + 02	1.787e + 03	3.649e + 02	1.563e + 03	3.595e + 02
50	75	2.130e + 03	3.272e + 02	2.095e + 03	3.432e + 02	2.002e + 03	3.541e + 02
75	100	9.385e + 02	1.997e + 02	1.173e + 03	2.375e + 02	1.340e + 03	2.683e + 02
100	125	1.972e + 02	7.824e + 01	3.372e + 02	1.075e + 02	4.852e + 02	1.350e + 02
125	150	1.938e + 01	2.135e + 01	5.006e + 01	3.438e + 01	9.822e + 01	4.854e + 01
900	925	6.237e − 03	9.366e − 02	1.524e − 02	1.691e − 01	5.042e − 02	3.020e − 01
925	950	3.956e − 01	5.737e + 00	2.430e + 00	1.498e + 01	1.120e + 01	3.118e + 01
950	975	3.391e + 01	7.092e + 01	1.243e + 02	1.418e + 02	3.513e + 02	2.393e + 02
975	1000	1.891e + 03	4.461e + 02	3.735e + 03	6.545e + 02	6.226e + 03	8.751e + 02
1000	1025	3.956e + 04	1.570e + 03	4.422e + 04	1.789e + 03	4.732e + 04	1.987e + 03
1025	1050	8.281e + 04	2.320e + 03	7.428e + 04	2.346e + 03	6.762e + 04	2.387e + 03
1050	1075	6.832e + 04	1.657e + 03	6.906e + 04	1.732e + 03	6.856e + 04	1.784e + 03
1075	1100	1.030e + 03	2.917e + 02	1.097e + 03	3.016e + 02	1.154e + 03	3.103e + 02
1100	1125	1.152e + 03	2.477e + 02	1.118e + 03	2.559e + 02	1.088e + 03	2.637e + 02
1125	1150	1.040e + 03	2.110e + 02	1.154e + 03	2.362e + 02	1.260e + 03	2.609e + 02
1150	1175	4.188e + 02	1.318e + 02	4.831e + 02	1.561e + 02	5.423e + 02	1.790e + 02
1175	1200	1.259e + 02	6.805e + 01	1.849e + 02	8.879e + 01	2.486e + 02	1.087e + 02

Further reading

The following books are suggested for further reading. The chapters to which they are particularly relevant are indicated by bracketed numbers at the end of the reference.

Andrews, D. G. (1999). *An Introduction to Atmospheric Physics*. Cambridge University Press. (1,2,3,4,5,7,8,11,12)

Andrews, D. G., Holton, J. R. & Leovy, C. B. (1987). *Middle Atmosphere Dynamics*. Academic Press. (5,8,10)

Gill, A. E. (1982). *Atmosphere–Ocean Dynamics*. Academic Press. (7,8,9,10)

Goody, R. M. (1995). *Principles of Atmospheric Physics and Chemistry*. Oxford University Press. (1,2,3,4,5,6,9)

Goody, R. M. & Yung, Y. L. (1989). *Atmospheric Radiation: Theoretical Basis*. 2nd edn. Oxford University Press. (1,2,4,6)

Green, J. (1999). *Atmospheric Dynamics*. 3rd edn. Cambridge University Press. (7,8,9,10,11)

Hartmann, D. L. (1994). *Global Physical Climatology*. Academic Press. (10,14)

Holton, J. R. (1992). *An Introduction to Dynamic Meteorology*. Academic Press. (7,8,9,10,11)

Houghton, J. T. (1997). *Global Warming: The Complete Briefing*. 2nd edn. Cambridge University Press. (14)

Houghton, J. T., Taylor, F. W. & Rodgers, C. D. (1984). *Remote Sounding of Atmospheres*. Cambridge University Press. (12)

IPCC (2001). *Climate Change 2001: The Scientific Basis*. Cambridge University Press.

Iribarne, J. V. & Godson, W. L. (1973). *Atmospheric Thermodynamics*. Reidel, Dordrecht. (3)

Jacobson, M. Z. (1999). *Fundamentals of Atmospheric Modelling*. Cambridge University Press. (1,2,3,4,5,6,7,9,11)

James, I. N. (1994). *Introduction to Circulating Atmospheres*. Cambridge University Press. (7,8,10,11)

Lorenz, E. N. (1993). *The Essence of Chaos*. UCL Press. (13)

Mason, B. J. (1974). *The Physics of Clouds*. Oxford University Press. (6)

Monteith, J. (1973). *Principles of Environmental Physics.* Edward Arnold. (9,14)

Richardson, L. F. (1922). *Weather Prediction by Numerical Processes.* Cambridge University Press (reprinted by Dover Publications, 1966). (11)

Salby, M. L. (1996). *Fundamentals of Atmospheric Physics.* Academic Press. (1,2,3,4,5,6,7,8,9,10)

Trenberth, K. E. (ed.). (1992). *Climate System Modelling.* Cambridge University Press. (11,14)

Wayne, R. P. (1991). *The Chemistry of Atmospheres.* 2nd edn. Oxford University Press. (5,14)

World Meteorological Organization (1994). *Scientific Assessment of Ozone Depletion: 1994.* WMO, Geneva. (5)

References to works cited in the text

Air Force Cambridge Research Laboratories (1965). *Handbook of Geophysics and Space Environments.*

Anderson, S. M. & Mauersberger, K. (1992). *Geophys. Res. Lett.*, **19**, 933–6.

Andrews, D. G. (1999). *An Introduction to Atmospheric Physics.* Cambridge University Press.

Andrews, D. G., Holton, J. R. & Leovy, C. B. (1987). *Middle Atmosphere Dynamics.* Academic Press.

Atkins, M. J. & Woodage, M. J. (1985). *Meteor. Mag.*, **114**, 227–33.

Baede, A. P. M. et al. (2001). In *Climate Change 2001: The Scientific Basis*, IPCC Cambridge University Press, ch. 1.

Batchelor, G. (1967). *An Introduction to Fluid Dynamics.* Cambridge University Press.

Berger, A. L. (1982). *Quaternary Res.*, **9**, 139–67.

Berger, A. (1995). *World Survey of Climatology* (ed. H. E. Landsberg), **16**, 21–69, Elsevier, Amsterdam.

Brewer, A. W. (1949). *Q.J.R. Met. Soc.*, **75**, 351–63.

Briggs, G. & Taylor, F. (1982). *Photographic Atlas of the Planets.* Cambridge University Press.

Brion, J. et al. (1993). *Chem. Phys. Lett.*, **213**, 610–12.

Broecker, W. S. & Denton, G. H. (1990). *Scien. Amer.*, **262**, 43–50.

Budyko, M. I. (1969). *Tellus*, **21**, 611–19.

Canby, T. Y. (1984). *Natn. Geogr. Mag.*, 144–83.

Chapman, S. (1930). *Mem. R. Met. Soc.*, **3**, 103.

Charney, J. G. (1973). In *Dynamical Meteorology*, ed. P. Morel. Reidel, Dordrecht, Holland.

Charney, J. G. & Drazin, P. G. (1961). *J. Geophys. Res.*, **66**, 83–109.

Charney, J., Fjörtoft, R. & von Neumann, J. (1950). *Tellus*, **2**, 237–54.

Clough, S. A., Grahame, N. S. & O'Neill, A. (1985). *Q.J.R. Met. Soc.*, **111**, 335–58.

COSPAR (1972). *COSPAR International Reference Atmosphere (CIRA).* Akademie-Verlag, Berlin.

COSPAR (1988). *COSPAR International Reference Atmosphere, Part I, Advances in Space Research*, **8**, nos 5 & 6.

COSPAR (1990). *COSPAR International Reference Atmosphere, Part II, Advances in Space Research*, **10**, no. 12.

Crane, A. J. (1977). Unpublished D.Phil. Thesis, University of Oxford.

Crutzen, P. J. (1971). *J. Geophys. Res.*, **76**, 7311–27.

Cullen, M. J. P. & Davies, T. (1991). *Q.J.R. Met. Soc.*, **117**, 993–1002.

Cullen, M. J. P. & Purser, R. J. (1984). *J. Atmos. Sci.*, **41**, 1477–97.

Cullen, M. J. P. & Roulstone, I. (1993). *J. Atmos. Sci.*, **50**, 328–32.

Curtis, P. D., Houghton, J. T., Peskett, G. D. & Rodgers, C. D. (1974). *Proc. R. Soc. Lond. A*, **337**, 135–50.

Ditchburn, R. W. & Young, P. A. (1962). *J. Atmos. & Terres. Phys.*, **24**, 127–39.

Dobson, G. M. B. (1914). *Q.J.R. Met. Soc.*, **40**, 123–35.

Dütsch, H. U. (1968). *Q.J.R. Met. Soc.*, **94**, 483–97.

Dütsch, H. U. (1971). *Adv. in Geophys.*, **15**, 219–322.

Dutton, J. A. & Johnson, D. R. (1967). *Adv. in Geophys.*, **12**, 333–436.

Eady, E. J. (1949). *Tellus*, **1**, 33–52.

Edwards, J. M. & Slingo, A. (1996). *Q.J.R. Met. Soc.*, **122**, 689–719.

Eisberg, R. M. (1961). *Fundamentals of Modern Physics*. John Wiley.

Eyre, J. R., Brownscombe, J. L. & Allam, R. J. (1984). *Meteor. Mag.*, **113**, 264–71.

Fjörtoft, R. (1953). *Tellus*, **5**, 225.

Fritts, D. C. (1984). *Rev. Geophys. & Space Phys.*, **22**, 275–308.

Gage, K. S. & Nostram, G. D. (1986). *J. Atmos. Sci.*, **43**, 729–39.

Gerbier, N. & Berenger, M. (1961). *Q.J.R. Met. Soc.*, **87**, 13–23.

Gill, A. E. (1982). *Atmosphere-Ocean Dynamics*. Academic Press.

Goldman, A. (1968). *J. Quant. Spectrosc. & Radiat. Transfer*, **8**, 829–31.

Goody, R. M. & Yung, Y. L. (1989). *Atmospheric Radiation: Theoretical Basis*. 2nd edn. Oxford University Press.

Goody, R. M. & Walker, J. C. G. (1972). *Atmospheres*. Prentice-Hall.

Gregory, D. & Rowntree, P. R. (1990). *Month. Weather. Rev.*, **118**, 1483–1506.

Hanel, R. A., Schlachman, B., Rodgers, D. & Vanous, D. (1971). *Appl. Opt.*, **10**, 1376–82.

Hansen, J. et al. (1992). *Geophys. Res. Lett.*, **19**, 215–18.

Harwood, R. S. (1975). *Q.J.R. Met. Soc.*, **101**, 75–94.

Harwood, R. S. & Pyle, J. A. (1975). *Q.J.R. Met. Soc.*, **101**, 723–48.

Hide, R. (1974). *Proc. R. Soc. London A*, **336**, 63–84.

Hide, R. (1982). *Q. Jl. R. Astr. Soc.*, **23**, 220–35.

Hide, R. & Mason, P. J. (1975). *Adv. in Phys.*, **24**, 47–100.

Hoskins, B. J. (1975). *J. Atmos. Sci.*, **32**, 233–42.

Hoskins, B. J., McIntyre, M. E. & Robertson, A. W. (1986). *Q.J.R. Met. Soc.*, **111**, 877–946.

Houghton, J. T. (1963). *Q.J.R. Met. Soc.*, **89**, 319–31.

Houghton, J. T. (1972). *Bull. Amer. Met. Soc.*, **53**, 27–8.

Houghton, J. T. (1978). *Quart. J. R. Met. Soc.*, **104**, 1–29.

Houghton, J. T. (1991). *Phil. Trans. R. Soc. Lond. A*, **337**, 521–72.

Houghton, J. T. (1997). *Global Warming: The Complete Briefing*. 2nd edn. Cambridge University Press.

Houghton, J. T. & Smith, S. D. (1966). *Infra-red Physics*. Oxford University Press.

Houghton, J. T., Taylor, F. W. & Rodgers, C. D. (1984). *Remote Sounding of Atmospheres*. Cambridge University Press.

Inn, E. C. Y. & Tanaka, Y. (1953). *J. Opt. Soc. Amer.*, **43**, 870.

Ingersoll, A. P. (1969). *J. Atmos. Sci.*, **26**, 1191–8.

IPCC (1990). *Climate Change, The IPCC Scientific Assessment* (eds J. T. Houghton, G. J. Jenkins & J. J. Ephraums). Cambridge University Press.

IPCC (1996). *The Science of Climate Change* (eds J. T. Houghton, L. G. Meira Filho, B. A. Callander, N. Harris, A. Kattenberg & K. Maskell). Cambridge University Press.

IPCC (2001). *Special Report on Emission Scenarios, Intergovernmental Panel on Climate Change*. Cambridge University Press.

IPCC (2001). *Climate Change 2001: The Scientific Basis*. Cambridge University Press.

Jeans, J. H. (1940). *Kinetic Theory of Gases*. Cambridge University Press.

Jeffreys, H. (1925). *Q.J.R. Met. Soc.*, **51**, 347–56.

Kennedy, J. S. (1964). *Energy Generation Through Radiation Processes in the Lower Stratosphere*. M.I.T. Report, no. 11, Boston.

Kiehl, J. T. & Trenberth, K. E. (1997). *Bull. Amer. Met. Soc.*, **78**, 197–208.

Kraichman, R. H. & Montgomery, D. (1979). *Rep. Prog. Phys.*, **43**, 547–619.

Krueger, A. J., Heath, D. E. & Mateer, C. L. (1973). *Pure and Appl. Geophys.*, **106–8**, 1254–63.

Kurucz, R. L. (1995). *Laboratory and Astronomical High Resolution Spectra*, Astron. Soc. of the Pacific Conf Series **81** (eds A. J. Sauval, R. Blomme & N. Grevesse), 17–31.

Leovy, C. B. (1964). *J. Atmos. Sci.*, **21**, 327–41.

Leovy, C. B. (1969). *Appl. Opt.*, **8**, 1279–86.

Lighthill, J. (1986). *Proc. R. Soc. Lond. A*, **407**, 35–50.

Lindzen, R. S. (1971). In *Mesospheric Models and Related Experiments* (ed. G. Fiocco. Reidel), Dordrecht, Holland.

Lorenc, A. C., Bell, R. S. & Macpherson, B. (1991). *Q.J.R. Met. Soc.*, **117**, 59–89.

Lorenz, E. (1955). *Tellus*, **7**, 157–69.

Lorenz, E. (1963). *J. Atmob. Sci.*, **20**, 130–41.

Lorenz, E. (1967). *The Nature and Theory of the General Circulation of the Atmosphere*. World Meteorological Organization, Geneva.

Lorenz, E. (1975). *Climate Predictability*, in GARP Publication Series no. 16, World Meteorological Organization, Geneva.

Lorenz, E. (1982). In *Problems and Prospects in Long and Medium Range Forecasting*, 1–20. European Centre for Medium Range Weather Forecasts, Reading.

McClatchey, R. A., Benedict, W. S., Clough, S. A., Burch, D. E., Calfee, R. F., Fox, K., Rothman, L. S. & Garing, J. S. (1973). *AFCRL Atmospheric Absorption Line Parameters Compilation*. Air Force Cambridge Research Laboratories Environment Research Papers no. 434.

McClatchey, R. A. & Selby, J. E. A. (1972). *Atmospheric Transmittance 7–30 μm*. Air Force Cambridge Research Laboratories Environmental Research Papers no. 419.

Malicet, J. et al. (1995). *J. Atm. Chem.*, **21**, 263–73.

Malkmus, W. (1967). *J. Opt. Soc. Amer.*, **57**, 323–9.

Manabe, S. & Wetherald, R. T. (1967). *J. Atmos. Sci.*, **24**, 241–59.

Matveev, L. T. (1967). *Fundamentals of General Meteorology: Physics of the Atmosphere.* Israel Programme for Scientific Translations, Jerusalem.

Milne, E. (1930). Reprinted in *Selected Papers on the Transfer of Radiation* (ed. D. H. Menzel). Dover, 1966.

Milton, S. F. & Wilson, C. A. (1996). *Month. Weath. Rev.*, **124**, 2023–45.

Mintz, Y. (1961). In *The Atmospheres of Mars & Venus.* Publication 944. Ad Hoc Panel on Planetary Atmosperes of Space Science Board, National Academy of Sciences, Washington D.C.

Murgatroyd, R. J. (1970). In *The Global Circulation of the Atmosphere* (ed. G. Corby). R. Met. Soc. London, 159–95.

Murgatroyd, R. J. & Singleton, F. (1961). *Q.J.R. Met. Soc.*, **87**, 125–35.

Newell, R. E., Kidson, J. W., Vincent, D. G. & Boer, G. J. (1972). *The General Circulation of the Tropical Atmosphere*, vol. 1. M.I.T. Press, Boston.

Newton, C. W. (1970). In *The Global Circulation of the Atmosphere* (ed. G. Corby). R. Met. Soc. London.

Oort, A. H. & Peixoto, J. P. (1974). *J. Geophys. Res.*, **18**, 2705–19.

Oort, A. H. & Piexoto, J. P. (1983). *Adv. in Geophys.*, **25**, 355–490.

Oort, A. H. & Vonder Haar, T. H. (1976). *J. Phys. Oceanog.*, **6**, 781–800.

Palmer, T. N. (1999). *J. Clim.*, **12**, 575–91.

Palmer, T. N. (2000). *Rep. Prog. Phys.*, **63**, 71–116.

Palmer, T. N., Schutts, G. J. & Swinbank, R. (1986). *Q.J.R. Met. Soc.*, **112**, 1001–39.

Penndorf, R. (1957). *J. Opt. Soc. Amer.*, **47**, 176.

Pivovonsky, M. & Nagel, M. R. (1961). *Table of Black-body Functions.* Macmillan.

Plumb, R. A. (1982). *Aust. Met. Mag.*, **30**, 107–21.

Prentice, I. C. et al. (2001). In *Climate Change 2001: The Scientific Basis*, IPCC. Cambridge University Press, ch. 3.

Rasool, S. I. & DeBergh, G. (1970). *Nature, Lond.*, **226**, 1037–9.

Raynaud, D. et al. (1993). *Science*, **259**, 926–34.

Read, P. L. & Hide, R. (1984). *Nature*, **308**, 45–8.

Richardson, L. F. (1922). *Weather Prediction by Numerical Processes.* Cambridge University Press (reprinted by Dover, 1966).

Schofield, J. T. & Taylor, F. W. (1983). *Q.J.R. Met. Soc.*, **109**, 57–80.

Schwalb, A. (1972). NOAA Technical Memorandum, Washington D.C.

Sellers, W. O. (1969). *J. Appl. Met.*, **8**, 392–400.

Smith, R. N. B. (1990). *Q.J.R. Met. Soc.*, **116**, 435–60.

Smith, W. L. (1976). *Proceedings of Study Conference on Four-dimensional Assimilation.* World Meteorological Organization, Geneva.

Starr, V. P. (1968). *Physics of Negative Viscosity Phenomena.* McGraw-Hill.

Stephens, G. L. & Webster, P. J. (1981). *J. Atmos. Sci.*, **38**, 235–47.

Thekaekara, M. P. (1973). *Solar Energy*, **14**, 109–27, Pergamon.

Taylor, F. W. et al. (1972). *Appl. Opt.*, **11**, 135–41.

Thomas, L. (1999). *Meteor. Appl.*, **6**, 133–42.

Uccellini, L. W., Keyser, D., Brill, K. F. & Wash, C. H. (1985). *Month. Weath. Rev.*, **113**, 962–88.

Vasavada, A. R. et al. (1998). *Icarus*, **135**, 265–75.

Vonder Haar, T. & Suomi, V. (1971). *J. Atmos. Sci.*, **28**, 305–14.

Vigroux, E. (1953). *Annales de Phys.*, **8**, 709.

Wayne, R. P. (1991). *The Chemistry of Atmospheres.* 2nd edn. Oxford University Press.

White, A. A. & Bromley, R. A. (1995). *Q.J.R. Met. Soc.*, **121**, 399–418.

Wiin-Nielsen (1981). *Atmospheric-Ocean*, **19**, 89–215.

Williams, A. P. (1971). Unpublished D.Phil. Thesis, University of Oxford.

World Meteorological Organization (1986). *WMO/NASA Ozone Assessment.* WMO, Geneva.

World Meteorological Organization (1994). *Scientific Assessment of Ozone Depletion: 1994.* WMO, Geneva.

Zhang, M. H., Hack, J. J., Kiehl, J. T. & Cess, R. D. (1994). *J. Geophys. Res.*, **99**, 5525–37.

Answers to problems
and hints to their solution

1.1 Earth: 276K,
   Venus: 326K,
   Mars: 224K,
   Jupiter: 121K.

1.2 $10.8\,W\,m^{-2}$.

1.3 $p(z) = p_0 \left(1 - \dfrac{\alpha z}{T_0}\right)^{Mg/\alpha R}$.

1.4 (1) 19.5 km; (2) 14.2 km (use the result of q 1.3).

1.5 3.07%: ~20%.

1.6 Mass of atmosphere $= 5.27 \times 10^{18}\,kg$.
   Thermal capacity of atmosphere $= 5.30 \times 10^{21}\,J\,K^{-1}$.
   Thermal capacity of the oceans $= 5.69 \times 10^{24}\,J\,K^{-1}$.

1.7 Using the value for c_p for CO_2 of $834\,J\,K^{-1}\,kg^{-1}$ ($\approx \frac{9}{2}R$) and for H_2 of $14450\,J\,K^{-1}\,kg^{-1}$ ($\approx \frac{7}{2}R$),
   Mars: 4.5
   Venus: 10.6
   Jupiter: $1.8\,K\,km^{-1}$.

1.8 $9.64\,K\,km^{-1}$ for damp air compared with $9.76\,K\,km^{-1}$ for dry air.

1.9 Solve equation iteratively or graphically: radius of balloon 13.3 m.

1.10 Period of oscillation = 9.8 minutes.

2.2 $\chi_0^* = 1.814$; $\phi = 183\,W\,m^{-2}$; Temperature discontinuity 20.7 K.

2.4 $\tau = 0.14$ days.

2.5 Consider rotation of planet around the sun, as well as about its own axis; 1 Venus solar day = 116 earth days, $\tau \approx 0.44$ Venus days.

2.6 $\chi_0^* \approx 224$.

3.1 $T^* = T(1 + m/\varepsilon)/(1 + m)$.

The required saturation mixing ratios may conveniently be read from the tephigram chart:

at 273 K, $T^* - T = 0.64$ K,

at 290 K, $T^* - T = 2.14$ K,

and

at 300 K, $T^* - T = 4.11$ K.

3.3 (i) 3.5, 4.2, 5.2.

(ii) 7.7 K km^{-1}.

3.7 (1) 30 kPa.

(2) Ascent is everywhere stable for dry air. It is stable for saturated air (if the mixing ratio is increased so that the air is saturated) above 78 kPa.

(3) At 100 kPa, 8.3 g kg^{-1}; at 50 kPa, 0.75 g kg^{-1}.

(4) 2 K.

(5) 95 kPa.

(6) 65 kPa.

3.9 Work done on 1st stage of ascent = 7 J.

Energy released in 2nd stage = 65 J.

3.10 The temperature of the final mixture is 18°C, and hence, for condensation to occur, mixing ratio of the mixture = 13.1 g kg^{-1}. Relative humidity = 93.7%.

3.11 (2) The approximation is that the process is replaced by one which is isentropic.

$\bar{\theta} = 28.6$°C.

$\bar{m} = 18.5$ g kg^{-1}.

Condensation level $\simeq$ 94 kPa.

3.12 Energy released = 6800 J kg^{-1}.

4.4 W m^{-2}:

Average solar input 245 W m^{-2}.

3.13 4×10^6 J m^{-2}.

4.1 At 0.3 μm, 0.637; at 0.6 μm, 0.061; at 1 μm, 0.008.

4.2 7.5%.

4.3 7.10^8 m^{-3}.

4.4 Turbidity coefficients, 0.74, 0.15.

Extinctions, 0.826, 0.304, 0.146.

4.6 $h_m = F_s e^{-1/3} g/2c_p p_m$, 0.88 K hr^{-1}. Note this is the heating rate with the sun overhead.

4.7 If $B_\nu(T)$ is proportional to solar flux, i.e. ignoring all dynamics, changes of ozone concentration etc., temperature difference is ~7 K.

4.8 (1) 0.568 kPa, (2) 2.13 kPa, (3) 0.133 kPa.

4.15 0.97 (0.98 for CO_2, 0.99 for H_2O);
0.68 (0.83 for CO_2, 0.82 for H_2O).

4.17 Mass mixing ratio of $H_2O = 0.622\,e/p$ (3.11).
(1) 0.924, (2) 0.92.
For last part use (4.20), multiply pathlength by 1.66 in transmission term. Assume spectral interval ~500 cm^{-1} wide. Answer 1.0 K day^{-1}.

4.19 To perform the integral over s write it as

$$\underset{\varepsilon \to 0}{\text{Lt}}\; N_0 \iint \int_\varepsilon^\infty \frac{1}{s}\exp\left(-\frac{s}{\sigma}\right)ds[1-\exp(-sf\rho l)]\,dv$$

$$= N_0 \int \left[\int_\varepsilon^\infty \frac{1}{s}\exp\left(-\frac{s}{\sigma}\right)ds - \int_{\varepsilon'}^\infty \frac{1}{s'}\exp\left(-\frac{s'}{\sigma}\right)ds'\right]dv$$

where $\varepsilon' = \varepsilon(1 + \sigma f\rho l)$.

5.1 Assume mixing ratio of He at 120 km same as at the surface (Appendix 3) and that oxygen completely dissociated at 120 km. For temperature of 800 K, 770 km; for 1000 K, 940 km; for 1400 K, 1260 km.

5.2 If T is orbital period, $\Delta T/T = -1.3 \times 10^{-7}$ per orbit.

5.3 8.3×10^{-10} m s^{-1}.

5.5 1.36×10^{11} m^{-3}; 2.2×10^{11} m^{-2}s^{-1}.

5.7 The resulting temperature profile rises sharply above 120 km, then flattens out so that above 400 km the temperature is substantially constant at about 1250 K.

5.8 Total potential energy, 130 J m^{-2}.
Power absorbed (from information in problem 5.7), 80 J m^{-2}. Thus we might expect $\Delta T/T$ ~0.6 compared with ~0.3 from fig. 5.4.

5.9 2.7 kPa.

5.11 Ratio about 1:100.

5.12 With tropopause temperature 196 K, volume mixing ratio 9.10^{-6}.

5.13 $(J_3 + k_2 n_2 m_2)^{-1}$; $\dfrac{J_3 + k_2 n_2 n_M}{4n_2 (J_2 J_3 k_2 k_3 n_M)^{1/2}}$.

First time constant: 0.76 s at 40 km; 0.05 s at 30 km; and 2×10^{-3} s at 20 km.
Second time constant: 1.6×10^5 s at 40 km; 1.3×10^6 s at 30 km; 1.3×10^7 s at 20 km.

5.14 0.4 K day^{-1}.

5.16 0.02.

5.17 $g_2/g_1 = 2$ for 15 μm CO_2 band; cooling rate $\simeq$ 5 K day^{-1}.

5.18 90 K day^{-1}.

5.19 Transmission at line centre $\simeq$ 0.66.

5.20 7.5×10^9; 2.2×10^{-5}.

6.3 (1) 26.5 min; (2) 7.34 hr.

6.4 91.9 days; 22 hours; 13.2 min.

6.5 22.9 min; −9.6°C.

6.6 $A = 2/3$, $\tau = 1/3$.

6.9 $A = 0.6635$, $\tau = 0.3353$, $(1 - A - \tau) = 5.97 \times 10^{-3}$.

6.10 19.56.

6.11 $\beta = 0.6138$, $\chi_0^* = 34.4$.

6.12 $\beta = 0.0266$, with $\chi_0^* = 4$, $\tau = 0.067$.

6.13 0.9972.

6.15 $T_c^4 = (1 - A)T_0^4$;
$T_1^4 = 2(1 - A)T_0^4$.

7.1 $\Omega^2 R = 3.457 \times 10^{-3} g$: 5.95 arc minutes.

7.3 Magnitudes of terms (cm s^{-2}):

$$\frac{du}{dt}, \frac{dv}{dt} \simeq 10^{-2}, \frac{uv\tan\phi}{a}, \frac{u^2\tan\phi}{a} \simeq 10^{-3},$$

$2\Omega u \sin\phi$, $2\Omega v \sin\phi \simeq 10^{-1}$, $\Omega w \cos\phi \simeq 10^{-4}$.
uw/a, $vw/a \simeq 10^{-6}$.

7.4 0.39 kPa/100 km.

7.5 7.29 m s^{-1}.

7.6 1 day at 30° latitude.

7.7 0.194. On Mars, 0.20; on Venus, 47.1.

7.8 ~10%.

7.11 $g = g_0 a^2 (a + z)^{-2}$.

7.12 19 m s^{-1}.

7.13 Consider pressures on either side of the frontal surface, p_1, p_2, and write

$$\delta p_1 = \frac{\partial p_1}{\partial x}\delta x + \frac{\partial p_1}{\partial z}\delta z$$

and similarly for δp_2. At front, $p_1 = p_2$, $\delta p_1 = \delta p_2$ and $\tan\alpha = \delta z/\delta x$. $\alpha = 0.41°$ at 30° latitude.

7.15 Temperature will fall.

7.16 $V_{20\text{kPa}} = 17.9 \text{ m s}^{-1}$ at 45° latitude.

7.17 Average temperature gradient, 1 K per 100 km; wind speed, 33.5 m s^{-1}.

7.25 Transformations of the form derived in q. 7.24 are applied to the continuity equation.

7.26 For scalar ϕ and vector **F**, curl ϕ**F** = ϕ curl **F** + $\nabla\phi \wedge$ **F**. Also curl $\nabla\phi = 0$, and g is the gradient of a scalar potential.

8.4 $1.88 \times 10^{-2} \text{s}^{-1}$.

8.5 $\lambda_h \simeq 10 \text{ km}$, $\lambda_v \simeq 1 \text{ km}$, horizontal phase speed $\simeq 16 \text{ m s}^{-1}$.

8.6 $\omega_a = N_B$ if there is $2.2\,\mathrm{K\,km^{-1}}$ temperature inversion.

8.8 The kinetic energy density is comprised of terms of form $\frac{1}{2}\rho u^2 : 0.14\,\mathrm{m\,s^{-1}}$ if $\overline{T} = 260\,\mathrm{K}$.

8.9 (1) $3.34\,\mathrm{km}$; (2) $4.78\,\mathrm{km}$.

8.10 $(v_g)_h = N_B \dfrac{m^2}{(m^2 + k^2)^{3/2}}, (v_g)_v \equiv \propto \dfrac{N_B km}{(m^2 + k^2)^{3/2}}$.

8.12 $13\,\mathrm{m\,s^{-1}}$.

8.13 $5.7\,\mathrm{m\,s^{-1}}$.

8.14 $v_g = \overline{u} + \dfrac{\beta(k^2 - l^2)}{(k^2 + l^2)}$.

8.15 0.02; 0.01.

8.16 ζ may be approximated to the vorticity of geostrophic wind to an approximation of order Ro.

8.18 $1.5 \times 10^{-5} : 3 \times 10^{-5}$.

8.19 The method is as for q 8.10 or 8.14.

8.20 The method is as for q 8.2 or 8.11.

9.3 The effect of buoyancy must be considered. We expect $K_E \simeq K$; K_Q will be different.

9.5 If $\zeta_g \simeq 10^{-5}\,\mathrm{s^{-1}}$, $H = 10\,\mathrm{km}$, $K = 10\,\mathrm{m^2\,s^{-1}}$, $w_d \simeq 7\,\mathrm{mm\,s^{-1}}$. Spin down time $\simeq 4$ days.

Decay time under damping from eddy viscosity but ignoring vertical motion and secondary circulation $\simeq H^2/K$, ~100 days.

10.1 (a) 2.94; (c) 0.337; (e) 0.032.

10.12 With $\lambda_m = 4000\,\mathrm{km}$, $H = 10\,\mathrm{km}$, $\partial \overline{u}/\partial z = 2\,\mathrm{m\,s^{-1}\,km^{-1}}$, $\Delta u = 20\,\mathrm{m\,s^{-1}}$, $(kc_i)^{-1} \simeq 2.10^5\,\mathrm{s}$.

10.15 $62.2\,\mathrm{m\,s^{-1}}$.

10.16 $2 \times 10^{-6} : 4\,\mathrm{m\,s^{-1}}$.

10.18 353, 475, 857, 1462, 2216 K; 200 days.

10.20 $u = \dfrac{-u_0[\exp(D/H) - \exp(z/H)]\sin(y/L)}{[\exp(D/H) - 1]}$.

$w = \dfrac{Lu_0[1 - \cos(y/L)]\exp(z/H)}{\alpha\beta H[\exp(D/H) - 1]}$.

11.3 1.9×10^{-3}.

11.6 18%.

12.1 (a) 0.4%; (b) 1%; (c) 3%.

12.2 Use method of q 4.11; ~1.8 K.

12.5 The following theorem concerning differentiation under the integral sign will be found useful:

if $F(\beta, \alpha) = \int_\alpha^\beta f(x, a)dx$

$$\frac{dF}{da} = f(\beta, \alpha)\frac{d\beta}{da} - f(\alpha, a)\frac{d\alpha}{da} + \int_\alpha^\beta \frac{\partial f(x, a)}{\partial a}.$$

12.6 (a) 5.8%; (b) 1.7%; (c) 0.4%.

12.8 0.25 K.

12.10 420 km.

12.11 Scattering coeff. ~4.10^{-10} cm^{-1}; absorption coeff. 10^{-7} cm^{-1}.

12.12 (a) 9.5 km; (b) 19.9 km.

12.13 Condition for ray to emerge is that horizontal ray distant r from centre of planet shall have radius of curvature $>r$. From information in problem 12.12 this condition is found to be

$$-\frac{dn}{dr} < \frac{n}{r}.$$

For Venus atmosphere, assuming temperature of ~410 K, highest pressure 300 kPa.

12.14 4.5×10^{-5} to 4.5×10^{-2} kPa.

13.2 Last part for the three cases
   (i) 11, 16, 23;
   (ii) 9, 16, 20;
   (iii) error 0.0001 at Y_4, then $n = 13, 18, 22$.

14.1 (a) 12 000 B.P., +7%, −7%.
        120 000 B.P., −4%, +4%.
    (b) 12 000 B.P., −1.7%, +7.8%.
        120 000 B.P., +1.4%, −6.1%.

14.2 About 4 years.

14.3 0.14% assuming albedo of 0.3.

14.4 About 270 K.

14.6 0.25 W m^{-2}; 0.15 K.

14.11 0.7 K.

Index

absolute momentum, 195
absolute vorticity, 118
absorption, and solar radiation, 40–4
acoustic cut-off frequency, 114
acoustic waves, 176–7
active remote sensing, 199
adiabatic lapse rate, 4–5, 6*f*, 7
Advanced Microwave Sounding Unit
    (AMSU), 211–12, 223*f*
Advanced Scatterometer (ASCAT), 223*f*
Advanced Very High Resolution
    Radiometer (AVHRR), 203–4, 206*f*,
    223*f*
airglow, 72
albedo, 2, 85–6
Along Track Scanning Radiometer
    (ATSR), 225
altitude, and atmospheric temperature
    structure, 209
Analysis Correction Scheme, 179
angular momentum, 150, 161–2
Antarctica, and ozone hole, 71
anticyclones, 97, 106*f*, 163*f*
anvil clouds, 36*f*
Atlantic Ocean, and thermo-haline
    circulation, 264. *See also* oceans; sea
    surface
atmosphere. *See also* general circulation;
    global observation; mesosphere;
    numerical modelling; stratosphere;
    troposphere; turbulence; waves
    adiabatic lapse rate, 4–5
    chaos and chaotic behavior, 229–31
    definition of, 1
    dynamics of, 93–102

    equilibrium temperatures, 1–3
    hydrostatic equation, 3–4
    predictability, 231–40
    Sandström's theorem, 5–7
atmospheric limb, 218
atmospheric path, transmission of, 44–5
available potential energy, 24–9, 30*f*

Back-Scatter Ultraviolet Spectrometer
    (BUV), 217
band models, of radiation transfer, 49–50
baroclinic instability, 147, 154–7
baroclinicity and baroclinicity vector,
    128
baroclinic models, 174–6
baroclinic waves, 145
barotropic instability, 153–4
barotropic model, 173–4
β-plane approximation, 118
biochemical feedbacks, and climatic
    change, 260–1
Bjerknes-Jeffreys theorem, 109
black-body radiation, 9–10
Boltzmann balance, 72–3
boundary conditions, and predictability,
    238–9
boundary layer, 134
Bouquet's law, 10
Boussinesq approximation, 120, 147–8
Bowen ratio, 142
Brewer, A. W., 72
Brewer-Dobson circulation, 72
Brunt-Vaisala frequency, 112–14
butterfly effect, and chaotic behavior,
    231

313

¬ll ___ 642-3122